公務員入職
能力傾向測試
熱門試題王

Fong Sir 著

序言

未來數年，退休公務員的人數將持續增加，到2023年更是公務員的退休高峰期。每年政府均透過CRE公務員綜合招聘考試來招募人才，為市民提供優質服務。有志加入政府公務員團隊的人士，是時候作好準備，增加獲聘的機會。

公務員綜合招聘考試包括三份45分鐘的試卷，分為「中文運用」、「英文運用」、「能力傾向測試」，另有「基本法及國安法測試」。

本書特色如下：

- 收錄超過200條例題及模擬試題，全面剖析能力傾向測試的考核要訣。

- 深入分析5大題型：中文演繹推理、英文演繹推理、數據運用、數序推理及圖表分析，箇中的重點概念、解題和答題技巧，讓考生舉一反三，遇到何種題目也應付自如。

- 搜羅熱門試題題型，助你熟習出題方向。

- 試題詳盡講解，方便考生了解自己不足，從而作針對性的練習。

目錄

目錄

CHAPTER THREE

Data Sufficiency Test

CHAPTER FOUR

Numerical Reasoning

CHAPTER FIVE

Interpretation of Tables and Graphs

PART

ONE

演繹推理

答題技巧

考生須根據短文的內容，選出一個或一組推論。考生須假定短文的內容都是正確，憑邏輯推論出最正確的答案。

1. 矛盾關係

「矛盾關係」是指兩個語句或命題之間不能同真（必有一假），也不能同假（必有一真）。不能同真，就是説當其中一個命題真時，另一個命題必假；不能同假，就是説當其中一個命題假時，另一個命題必真。

例題：

如果生產下降或浪費嚴重，那麼將造成物資匱乏。如果物資匱乏，那麼或者物價暴漲，或者人民生活貧困。如果人民生活貧困，政府將失去民心。事實上物價沒有暴漲，而且政府贏得了民心。由此可見：

A. 生產下降，但是沒有浪費嚴重。

B. 生產沒有下降，但是浪費嚴重。

C. 生產下降，且浪費嚴重。

D. 生產沒有下降，且沒有浪費嚴重。

答案：D。

「政府贏得了民心」是對「如果人民生活貧困，政府將失去民心」這一充分條件假言命題的後件的否定，根據有效推理規則，可以推出否定的前件，即人民生活不貧困。

「人民生活不貧困」與「物價沒有暴漲」結合形成一個聯言命題，是對「如果物資匱乏，那麼或者物價暴漲，或者人民生活貧困」的後件的否定，根據有效推理規則，可以推出否定的前件，即物資不匱乏。

「物資不匱乏」是對「如果生產下降或浪費嚴重，那麼將造成物資匱乏」的後件的否定，根據有效推理規則，可推出否定的前件，即「生產沒有下降且浪費不嚴重」，故答案選D。

2. 三段論

「三段論」是指由三個命題構成的推理。具體説來，三段論是由包含著一個共同因素(邏輯中介)的兩個命題推出一個新的命題的推理。其基本規則是：

1.「所有A是B」+「所有B是C」，推出「所有A是C」。

2.「所有A是B」+「所有B不是C」，推出「所有A不是C」。

3.「有些A是B」+「所有B是C」，推出「有些A是C」。

4.「有些A是B」+「所有B不是C」，推出「有些A不是C」。

以上四條，是三段論推理的基本規則，我們要在上面的基礎上保證推理的正確性。同時，我們還要做到要熟知一下幾點：

1.「有些A是B」＝「有些B是A」

例題：所有參加此次運動會的選手都是身體強壯的運動員，所有身體強壯的運動員都是極少生病的，但是有一些身體不適的選手參加了此次運動會。

以下哪項不能從上述前提中得出？

A. 有些身體不適的選手是極少生病的

B. 有些極少生病的選手感到身體不適

C. 參加此次運動會的選手都是極少生病的

D. 極少生病的選手都參加了此次運動會

答案：D。

選項中A、B剛好符合「有些A是B」＝「有些B是A」，也就是這兩個選項是同義的，那麼這兩個選項肯定都是對的。而「所有A是B」不能推出「所有B是A」，那麼可見，D是一定錯的。那麼大家，知道這兩個特徵之後，答案就很容易得到。

2.「某個A不是B」，推出「有些A不是B」

例題：有些內地電視節目是反貪腐題材的，所有電視節目都具有看點，電視節目《小棉襖》不是反貪腐題材的，所有反覆題材的節目都受到了好評。

如果以上為真，以下各項不能確定真假的是：

A. 有些電視節目不是反覆題材的

B.《小棉襖》沒有收到好評

C. 有些電視節目受到了好評

D. 有些反貪腐題材的節目具有看點

答案：B。

利用「某個A不是B「可以推出「有些A不是B」，所以B項一定正確；利用「有些電視節目是反腐題材的」+「所有反覆題材的節目都受到了好評」，可以推出「有些霜視節目受到了好評」；「有些電視節目是反貪腐題材的，所有電視節目都具有看點」可以推出「有些反貪腐題材的節目具有看點」。只有B項無法推出。

3. 削弱論證法

削弱論證就是從選項中找一個能夠對題中論點構成質疑，使題目中的論點不成立的選項。在削弱論證中主要有以下幾種加強技巧。

第一種是「削弱論點」。所謂的削弱論點就是直接否定論點，其削弱力度是最強的。

第二種方法是「拆橋」，通俗的説就是斷開論據和論點之間的聯繫，對我們的論點就更起到了削弱作用。例如：論點是小明是好人，論據是小明每天都扶老婆婆過馬路。拆橋就是扶老婆婆過馬路的不一定是好人。

第三種方式是「削弱論據」，也就是增加一個反向的論據，通過新增加反向論據對題中論點構成削弱。

第四種就是「削弱前提」，也叫做「否定前提」。

在答題的過程中，如果讓我們選擇最有力削弱以上論點的是，如果這四種削弱方式同時出現，我們應該怎麼選擇？在這四種方法中削弱論點的力度是最強的，其次是否定前提和拆橋，最弱的是削弱論據。

下面來看一條題目：

研究發現，試管嬰兒的出生缺陷率約為9%，自然受孕嬰兒的出生缺陷率約為6.6%。這兩部分嬰兒的眼部缺陷比例分別為0.3%和0.2%，心臟異常比例分別為5%和3%，生殖系統缺陷的比例分別為1.5%和1%。因而可以說明，試管嬰兒技術導致試管嬰兒比自然受孕嬰兒出生缺陷率高。

以下哪項如果為真，最能質疑該結論？

A. 試管嬰兒要經過體外受精和胚胎移植過程，人為操作都會加大受精卵受損的風險。

B. 選擇試管嬰兒技術的父母大都有生殖系統功能異常，這些異常會令此技術失敗率增加。

C. 試管嬰兒在體外受精階段可以產生很多受精卵，只有最優質的才被揀選到母體進行孕育。

D. 試管嬰兒的父母比自然受孕嬰兒的父母年齡大很多，父母年齡越大，新生兒出生缺陷率越高。

答案：D。

題幹中的論點是「試管嬰兒技術導致試管嬰兒比自然受孕嬰兒出生缺陷率高」。

A項：「人為操作都會加大受精卵受損的風險」，舉例子說明試管嬰兒技術會導致嬰兒出生缺陷率高，加強作用，排除。

B項：說明試管嬰兒技術失敗率增加，論點是嬰兒出生缺陷率高，排除。

C項：屬於另有他因，說明「試管嬰兒比自然受孕嬰兒出生缺陷率高」的不是「試管嬰兒技術」的原因，而是「父母年齡大」。

所以，本題的答案為D。

4. 尋找題幹成立的大前提

在邏輯判斷中，加強削弱絕對是考察的重點，並且題目數量佔的比例也比較大，在論證類題目中，有這樣一種特殊的提問方式：「能成為上述結論隱含前提/假設/條件的是哪項？」

這類題目要求，對於題幹而言，除了要選擇有加強作用的選項，而且特殊之處在於不僅可以加強題幹，而且對於題幹而言是必不可少的一項。

對於必要條件選項的識別，往往可以採用否定代入的形式來選擇。也就是說在論證題當中，論點和前提存在一個邏輯關係，就是論點 前提，即：假設我們否定帶入A選項，發現原題當中的論點不成立，那麼A選項就是這個論點的前提，滿足否後必否前；假設我們否定帶入A選項，發現原題當中的論點沒有受到絲毫的影響，依然是成立的，那麼A選項就不是這道題的前提，因為不滿足否後否前，那麼接下來我們通過幾道題來練習一下。

例題：

近來網上出現了一則有關「公務員綠色出行」的報道，即A市教育局通過添置公務自行車，使得較近距離的公務活動不再派汽車，從而大幅降低了公務汽車使用率。B市環保局負責人看了這則報道，認為引進公務自行車能縮減財政開支，於是決定也採取這種做法。

以下哪項如果為真，不能構成上述決定的前提條件？

A. 自行車出行的方式結合B城市的地理環境以及道路交通狀況是可行又可取的

B. B市環保局的辦公人員規模和已有公車數量與A市教育局相近

C. 較近距離的公務活動有自行車取代汽車不會影響辦公效率

D. B市環保局原來的公務活動主要採取汽車出行的方式

答案：B。

論點：引進公務自行車能縮減B市財政開支。

論據：A市教育局通過添置公務自行車，使得較近距離的公務活動不再派汽車，從而大幅降低了公務汽車使用率。

題目要求選擇不能構成上述決定的前提條件，只需要把可以加強的選項排除掉即可。否定代入法，

A項：自行車出行的方式結合B城市的地理環境以及道路交通狀況如果不可行又不可取，題幹論點無法成立，為前提條件。

B項：B市環保局的辦公人員規模和已有公車數量與A市教育局是否相近不影響論點，不是決定因素。

C項：較近距離的公務活動有自行車取代汽車會影響辦公效率，題幹論點無法成立，為前提條件。

D項：B市環保局原來的公務活動主要不以採取汽車出行的方式，題幹論點無法成立，為前提條件。

故正確答案為B。

5. 排除法

排除法即按照順序讀題幹中信息，用一定的邏輯推理排除與題幹中所給信息不符合的選項，從而得出正確選項的一種方法。即按順序讀題，讀一句排一句。做條件確定的分析推理題時，最好優先使用排除法。

例題：

一次聚會上，米高遇到了湯姆、卡爾和喬治三個人，他想知道他們三人分別是幹什麼的，但三人只提供了以下信息：三人中一位是律師、一位是推銷員、一位是醫生；喬治比醫生年齡大，湯姆和推銷員不同歲數，推銷員比卡爾年齡小。

根據上述信息米高可以推出的結論是：

A. 湯姆是律師，卡爾是推銷員，喬治是醫生。

B. 湯姆是推銷員，卡爾是醫生，喬治是律師。

C. 湯姆是醫生，卡爾是律師，喬治是推銷員。

D. 湯姆是醫生，卡爾是推銷員，喬治是律師。

答案：C。

該題屬於條件確定的分析推理型題型，先用排除法去解答：

喬治比醫生年齡大，説明喬治不是醫生，排除A項。

湯姆和推銷員不同歲，説明湯姆不是推銷員，排除B項。

推銷員比卡爾年齡小，説明卡爾不是推銷員，排除D項。

故正確答案為C 。

試題練習

1. 已知有甲、乙、丙和丁四個數，甲、乙的和是大於丙、丁之和，甲、丁之和大於乙、丙之和，乙、丁之和大於甲、丙之和。請問這四個數中哪個最小？

A. 甲

B. 丙

C. 乙

D. 丁

2. 在一列火車上，來自英、意、日、德四國的**A**、**B**、**C**、**D**四位旅客，恰好相聚在某個車廂中。他們每人除了會說自己本國語言外，還會說其他三國語言中的一種，有一種語言三個人都會說。

這四位旅客交談的有關情況如下：

(1)B不會說英語，當**A**與**C**交談時，**B**卻能替二人作翻譯；
(2)A是日本人，**D**不會說日語，但他倆卻能毫無困難地交談；
(3)B、**C**、**D**三人找不到一種共同的語言進行交談；
(4)在四人之中，沒有一人既能用日語交談，又能用意大利語交談。

據此，可以推出三個人都會説的語言是：

A. 日語

B. 德語

C. 英語

D. 意大利語

3. 根據某國衛生部門的統計數據，貧窮家庭兒童的肥胖率要比富裕家庭兒童高出近一倍。根據該國政府最新公布的數據，全國11歲的貧窮家庭兒童當中，有超過24.3%的人出現體重超標的問題，而富裕家庭的同齡兒童肥胖率只有13.7%。因此專家認為，越貧窮的家庭，越有可能出現孩子體重超標的現象。

以下哪項如果為真，最能支持上述結論？

A. 貧窮家庭一般選擇到廉價超市購買食材

B. 貧窮家庭日常食用的，多是價格低廉但熱量高的食品。

C. 貧窮家庭兒童與富裕家庭兒童，喜歡的體育運動項目不同。

D. 貧窮家庭的孩子與富裕家庭的孩子，往往在不同的學校就讀。

4. 據某一年的調查報告顯示，75%的重大刑事案件未成年罪犯多是來自單親家庭。這說明，單親家庭將極大增加未成年人犯罪。

以下各項如果為真，最能削弱上述結論的是：

A. 這一年，許多家庭父母離婚是因為孩子是罪犯。

B. 這一年的罪犯比上一年增加了70%

C. 許多未成年罪犯的父母關係很好，對孩子極其關愛。

D. 總犯罪量不變，這一年成年人犯罪的案件數有驚人的增長。

5. 調查顯示，許多尋求醫療幫助的人處於很大的壓力下。醫學研究同時顯示，壓力能夠嚴重影響一個人抵抗感染的免疫系統。因此，當一個人處於壓力線的時候，他更可能得病。

以下選項最能支持上面的結論是：

A. 在放假期間，大學醫院所處理的疾病數目顯著增多。

B. 在考試期間，大學醫院所處理的疾病數目顯著減少。

C. 許多企業報道在職業感到管理壓力期間，缺席率顯著地減少。

D. 許多為職業提供健康保險的企業，同時提供應付壓力的研習班。

6. 某開發區發展委員會召開環境工作專題圓桌會議，參加會議的有委員會主任和副主任，以及委員會所屬的開發區環保局、工業局和農業局的局長和副局長。他們八個人均匀地坐在一張會議圓桌旁，只有一個同部門的正職和副職的座位被分隔開了。並且：

(1)委員會副主任對面的人是坐在環保局局長左邊的一位局長；

(2)工業局副局長左邊的人是坐在農業局局長對面的一位副局長；

(3)農業局局長右邊的人是一位副局長，這位副局長坐在委員會主任左邊第二個位置上的副局長的對面。

根據以上幾個條件，座位一定被割開的是：

A. 環保局的局長和副局長

B. 工業局的局長和副局長

C. 農業局的局長和副局長

D. 委員會的主任和副主任

7. **在一次介紹英國歷史的講座中，演講嘉賓介紹：根據1701年制定的英國王位繼承法，君主的長子擁有絕對的王位繼承優先權。現場聽眾中有人提出質疑：那不正確，包括現任女王在內，英國歷史上有很多女王都繼承了王位。這位現場聽眾可能產生了以下誤解，除了：**

A. 只在君主的長子才能夠合法繼承王位

B. 英國歷屆女王都作為第一順序繼承人直接繼承了王位

C. 英國歷史上諸位女王，可能並不都是通過王位繼承法合法繼承王位的。

D. 只有兒子才是合法的王位繼承人

8. **對於大部分東西來説，溫度越高，溶解的越快，可是對於食鹽來説，溫度升高了，溶解度並沒有多大的變化，因此：**

A. 用熱水溶解食鹽是明智之舉

B. 只有攪拌才能使食鹽更快地溶化

C. 用熱水溶解食鹽並非高人一等

D. 水溫對食鹽的溶解是至關重要的

9. 鋼琴考試結束後，關於甲、乙二人是否能通過考試，宿舍同學有如下討論：

甲說：或者我能通過，或者乙能通過，或者我和乙都能通過。

丙說：甲一定能通過。

丁說：乙能通過。

戊說：甲沒有可能通過。

考試成績出來後，證明四人的話有兩人說的是真的，兩人說的是假的。

根據以上陳述，可推出以下哪項？

A. 甲、乙都通過了

B. 甲、乙都未能通過

C. 甲通過了，乙未能通過。

D. 乙通過了，甲未能通過。

10. 某地方有南、北兩個綠化區，擬在其中種植一些果樹。員工們推薦了4種果樹備選：杏樹、桃樹、蘋果、柿。根據實際情況，還需要滿足以下3項種植要求：

(1)每區種植3種果樹；

(2)至少要在一個綠化區同時種植杏樹和蘋果；

(3)種植桃樹的綠化區也要種柿。

下列選項中，一定錯誤的是：

A. 兩個區都種有桃樹

B. 兩個區都種有蘋果

C. 只有一個區種有杏樹

D. 只有一個區種有蘋果

11. **現在越來越多的年輕人未註冊結婚就共同生活在一起了，結果由於種種原因，當發生財產等民事糾紛時，不能得到法律上的承認和保護。因此，註冊結婚有避免兩人感情關係破裂後，導致很多無謂糾紛的作用。**

以下哪一項最能加強上述論點？

A. 現代的婚姻越來越不穩定，離婚率越來越高。

B. 並非每對非婚同居者都會發生民事糾紛

C. 現代的年輕人更看重男女間的感情

D. 有些非婚同居者後來也去領了結婚證

12. **某商店被盜，警方掌握的線索如下：**

(1)甲、乙和丙三人中，最多有一個是罪犯。
(2)如果甲是罪犯，乙肯定是共犯。
(3)盜竊案發生時，丙正在家裡吃飯。
由以上三個條件，可以推論出：

A. 甲和丙都不是罪犯

B. 甲是罪犯

C. 乙是罪犯

D. 丙是罪犯

13. 某停車場為增加收入，打算改變其計時收費方法。之前的收費方法是在首兩小時，或者不到兩小時的時間內收取2元，而之後每小時收1元，而現在的收費方法是在首四小時或不到四小時的時間內收取4元，而往後每小時收取1元。

下面哪種説法如果正確，表明該計劃可以成功地增加收入？

A. 把車停在這個停車場內的人，通常把車停在「按天收費」區而非「按時收費」的停車區內。

B. 用來運營停車場的資金很大一部分被用來維護設備，而不是支付停車場職工工資。

C. 很少有人會在這個停車場內一次停車超過兩小時

D. 在過去的幾年內，停車場運營計時停車設備的成本，要高於從中獲得的利潤。

14. 甲、乙、丙三人各自舉著紅旗、綠旗和黃旗，分別從東面、南面和西面三個方向朝山頂攀登。甲不舉紅旗，也不從東面上山，舉紅旗的人從西面上山；乙舉著綠旗。由此可以推出：

A. 舉綠旗者從南面上山

B. 丙從東面上山

C. 舉黃旗的不是甲

D. 乙不從南面上山

15. 在南極海域冰冷的海水中，有一種獨特的魚類，牠們的血液和體液中具有一種防凍蛋白，因為該蛋白它們才得以存活並演化至今。但時至今日，該種魚類的生存卻面臨巨大挑戰。有人認為這是海水升溫導致的。

以下哪項如果為真，最能支持上述觀點？

A. 南極海水中的含氧量會隨氣溫上升而下降，缺氧導致防凍蛋白變性，易積於血管，導致供血不足，從而縮短魚的壽命。

B. 防凍蛋白能夠防止水分子凝結，從而保證南極魚類正常的活動，氣候變暖使得該蛋白變得可有可無。

C. 南極魚類在低溫穩定的海水中能夠持續地演化，而溫暖的海水不利於南極魚類的多樣性。

D. 並非所有南極物種都具有防凍蛋白，某些生活於副極地的物種並沒有這種蛋白。

16. **疾病預防控制中心的研究報告認為，煙草使用是首要的可預防死因。每年全世界有500萬人死於煙草相關疾病，其中每年死於二手煙的人數也高達60萬，四分之一以上是兒童。**

以下各項如果為真，哪項最不能質疑上述觀點？

A. 每年全世界有500多萬人，死於因為運動不足而引發的疾病。

B. 每年全世界有500多萬人，死於因為生活不規律而引發的疾病。

C. 每年全世界不低於1,405萬人的可預防死因，並非與煙草相關疾病。

D. 每年全世界因心理調試不當引發的疾病，成為人類首要的可預防死因。

17. **某食品包裝上標明：本食品可放心食用，因為其中不含防腐劑和人工色素。該食品包裝上所標明內容最可能假設了以下哪項？**

A. 有些食品中含防腐劑和人工色素

B. 安全的食品一定不含防腐劑和人工色素

C. 不含防腐劑和人工色素的食品一定是安全食品

D. 這些食品的生產者抓住了消費者的心理需求

18. **一項研究表明，那些在舒適環境裡工作的人，比在不舒適環境裡工作的人生產效率高25%。評價工作績效的客觀標準包括承辦工件數和工件的複雜程度。這表明：日益改善的工作環境可以提高工人的生產效率。**

以下哪項為真，最能削弱上述結論？

A. 平均來説，生產效率低的員工每天在工作場所的時間，比生產效率高的員工要少。

B. 舒適的環境比不舒適的環境，更能激勵員工努力工作。

C. 生產效率高的員工，通常得到舒適的工作環境作為回報。

D. 生產效率高的員工，不會比生產效率低的員工工時長。

19. **一位社會學家對已婚者和離婚者展開的調查發現：戀愛時間越長的夫妻，婚後鬧分手的機會率，會比那些戀愛時間較短的夫妻低。於是有人得出結論：要想以後的婚姻更穩固，必須延長戀愛時間。**

下列哪項為真，最能削弱上述結論？

A. 戀愛談得越長，兩人結婚的可能性會越低。

B. 情侶在戀愛期是否深入了解彼此，比戀愛時間長短更重要。

C. 「一見鍾情式」的速成婚姻容易破碎。

D. 戀愛時間長的人，往往會忘記戀愛的目的是婚姻。

20. 公司決定在王、陳、周、李、林、胡等6人中，挑選幾人去執行一項重要任務，執行任務的人選應滿足以下所有條件：

王、李兩人中只要一人參加；
李、周兩人中也只要一人參加；
王、陳兩人至少有一人參加；
王、林、胡3人中應有兩人參加；
陳和周要麼都參加，要麼都不參加；
如果林參加，李一定要參加。
據此，可以推出：

A. 王、陳不參加

B. 林、胡不參加

C. 周、李不參加

D. 李、林不參加

21. **某校最近進行了一項有關獎學金，對學習效率是否有促進作用的調整，調查內容包括自習的出勤率、完成作業所需要的平均時間、日平均閱讀量三項指標。**

結果表明：獲得獎學金的學生比那些沒有獲得獎學金的學生的學習效率要高出25%，由此得出結論：獎學金對幫助學生提高學習效率的作用是很明顯的。

以下哪項最能削弱以上結論？

A. 獲得獎學金的那些同學，通常是因為有好的學習習慣和高的學習效率。

B. 獲得獎學金的同學，可以更容易改善學習環境來提高學習效率。

C. 學習效率低的同學，通常學習時間長而缺少正常的休息。

D. 對學習效率高低與獎學金多少的關係的研究，應當採取定量方法進行。

22. 海豚是體型較小的鯨類，共62種，分布於世界各大洋，海豚一般體長1.2米至10米，體重45千克至9000千克，主要以小魚、烏賊、蝦、蟹為食。他們喜歡過「集體」生活，少則幾條，多則幾百條，海豚是一種本領超群、聰明伶俐的海中哺乳動物。

根據以上陳述，可以得出以下哪項結論？

A. 海豚是一種體型較小的哺乳動物

B. 有些聰明的哺乳動物生活在海裡

C. 海洋中的哺乳動物至少有62類

D. 海豚善解人意，能模仿人的動作

23. 調查發現，養寵物的人往往身體比較健康，心理比較樂觀。從以上的事例中，我們可以得出如下結論：

A. 養寵物會讓人身心健康

B. 身心健康的人才有精力養寵物

C. 對動物有愛心使人身心比較健康

D. 養寵物的人身心比較健康

24. 所有睡蓮都是晚上開的，這株蓮花是白天開的，所以這株蓮花不是睡蓮。上述推理的形式結構與以下哪項最為類似？

A. 所有職業短跑運動員都穿釘鞋，李先生不是職業運動員，所以李先生不穿釘鞋。

B. 所有山羊都長有鬍鬚，這隻羊沒有鬍鬚，所以這不是山羊。

C. 所有醫生都在醫院工作，王先生在醫院工作，所以他是醫生。

D. 所有納稅人都有存款，這位太太有存款，所以這位太太是納稅人。

25. 科學家對發掘於埃塞俄比亞哈達爾遺址的南方古猿足骨化石的第4根跖骨進行分析研究後發現，非洲南方古猿在當時便具有了定型的弓形足。他們據此認為，人類的祖先早在320萬年前就開始像現代人一樣用雙腳行走。

以下哪項為真，最能支持上述論證？

A. 只有分析第4根跖骨化石，才能發現非洲南方古猿具有定型的弓形足。

B. 只有南方古猿才是人類的祖先

C. 只有具有定型的弓形足，才能使用雙腳行走。

D. 只有使用雙腳行走，才具有定型的弓形足。

26. **在接受測謊器測驗時，只要服用少量某種常用鎮靜劑，就可使人在説謊時不被察覺。因為由測謊器測出的心理壓力反應可以通過藥物抑制，而不會產生顯著的副作用。由此推之，這種藥物對降低日常生活中的心理壓力也會很有效。**

此段話的假設前提是：

A. 鎮靜劑對於治療心理壓力有效

B. 對於心理壓力反應的抑制，增加了主觀的心理壓力。

C. 由測謊器測出的心理壓力，與日常生活的心理壓力相似。

D. 在測謊器測驗中説謊的人，經常顯示出心理壓力的跡象。

27. **十六世紀英國市場上流通的是金屬鑄幣，時間長了，人們發現足值與不足值的鑄幣可以一樣使用，於是，人們就把成色的足值貨幣(良幣)儲藏起來，而把不足值的貨幣(劣幣)趕緊花出去，結果，劣幣把良幣趕出了市場。這就是格雷欣法則，是一種「優敗劣勝」的自然與社會現象。**

以下哪種現象如果為真，最符合以上的結論？

A. 猴群中強者為王

B. 山寨手機在手機市場上流行一時

C. 優秀的文學作品大多是能夠廣為流傳的

D. 學校裡品學兼優的學生，總是容易脱穎而出。

28. **經A省的防疫部門檢測，在該省境內接受檢疫的水獺中，有2%感染上了瘋狗症。但是只有與人及其寵物有過接觸的水獺才接受檢疫。防疫部門的專家因此推測，該省水獺中感染有瘋狗症的比例，將大大小於2%。**

以下哪項如果為真，將最有利地支持專家的推測？

A. 在A省境內，與人及其寵物有接觸的水獺，只佔水獺總數的不到10%。

B. 在A省，感染有瘋狗症的寵物，約佔寵物總數的0.1%。

C. 在與A省毗鄰的B省境內，至今沒有關於水獺感染瘋狗症的疫情報告。

D. 與健康的水獺相比，感染有瘋狗症的水獺，更願意與人及其寵物接觸。

29. **某家電公司屬下有甲、乙、丙三個工廠：甲廠擅長生產雪櫃、洗衣機和微波爐；乙廠擅長生產洗衣機、冷氣機和消毒櫃；丙廠擅長生產冷氣機和消毒櫃。該家電公司經調查發現，如果兩個公司生產同樣的產品，一方面規模不夠經濟，另一方面會產生內部惡性競爭，而且一個廠生產三種產品，在人力和設備上也有問題，為更好地發揮各廠的優勢，公司的高層召集三個工廠的負責人對各自的產品作出協調，並作出了滿意的決策。**

以下哪項最可能是這幾個廠的產品選擇方案？

A. 甲廠生產雪櫃和消毒櫃，乙廠只生產洗衣機。

B. 甲廠生產雪櫃和洗衣機，乙廠生產冷氣機和消毒櫃。

C. 乙廠生產洗衣機和消毒櫃，丙廠生產冷氣機和微波爐。

D. 乙廠只生產洗衣機，丙廠生產冷氣機和消毒櫃。

30. 研究人員把一對橡膠手套進行消毒處理，然後把其中一隻手套放入細菌培養基，使其外表沾滿了大腸杆菌。試驗人員戴著這兩隻手套進行了一系列的打招呼動作，包括握手、碰拳和擊掌。研究表明，握手傳播的細菌數量相當於碰拳的十倍，而擊掌傳播的細菌數量居於兩者之間。因此，專家呼籲，尤其是在流感高峰季節，人們應該用碰拳問候來取代握手。

以下哪一項為真，最能反駁專家的觀點？

A. 試驗人員在進行打招呼的動作時，嘗試了不同力度均是相同結果。

B. 大腸杆菌是平時人們手上經常沾染的細菌類型

C. 流感致病細菌在握手時的傳播規律與大腸杆菌不同

D. 在某些地區，碰拳並不是特別流行的問候方式。

答案及解析

1. 答案：B。
根據題幹的意思可知：
(1)甲+乙>丙+丁
(2)甲+丁>乙+丙
(3)乙+丁>甲+丙
將以上三個觀察進行相應簡化：
(1)+(2)可得甲>丙
(2)+(3)可得丁>丙
(1)+(3)可得乙>丙，因此丙最小。

2.答案：B。
第一步：通過(1)和(3)判斷有共同語言的人
由(1)可知，C和B有共同語言，C和A卻沒有共同語言。由「C和B有共同語言」和(3)可知，C和D沒有共同語言，又「C和A沒有共同語言」，即可推出有共同語言的三人為A、B、D。
第二步：判斷會說的語言
由(4)可知，日本人(即A)不會意大利語，意大利人不會日語。由(2)可知，A、D交談的語言是德語或者英語，又由(1)可知，B不會說英語。所以三人都會說的語言是德語，故正確答案是B。

3. 答案：B。
由於A、C和D都是無關選項，與體重超標沒有必然的關聯。B屬於加強論據。

4. 答案：A。
以下是各個條件的分析：
A項：「未成年罪犯」導致「單親家庭」增多，與題幹結論為因果倒置，故完全削弱結論。
B項：罪犯的數目比上一年增加，但無法確定未成年罪犯所佔比重，無法斷定是削弱還是支持。
C項：父母關係好，對孩子極其關愛，並不是因為單親家庭才促使未成年人犯罪，削弱結論。
D項：選項「成年人犯罪」與「未成年人犯罪」，並不涉及「重大刑事案件」和「單親家庭」的結論，與結論無關。

5.答案：D。

第一步：找到論點和論據

本題論點是當一個人處於壓力線的時候，他更可能得病，論據是許多尋求醫療幫助的人處於很大的壓力下，壓力能夠嚴重影響一個人防被感染的免疫系統。

第二步：判斷加強方式

論點強調得病，論據強調壓力影響免疫，所以論點與論據間不存在明顯因果關係，需加強論證，在論點和論據間搭橋。

第三步：逐一判斷選項的作用

A項：放假期間疾病數目顯著增多，説明壓力少時得病增多，是削弱題幹的結論。

B項：考試期間疾病數目顯著減少，説明壓力大時得病減少，是削弱題幹的結論。

C項：職業感到管理壓力期間，缺席情況顯著地減少，缺席與得病無直接關係，屬無關選項。

D項：既提供健康保險又提供應付壓力的研習班，説明健康和壓力相關，即在得病和壓力影響免疫間搭橋，加強論證，是答案。

綜合上述各點，故正確答案為D。

6.答案：B。

根據題幹信息和條件，我們會發現只有工業局長和工業副局長是不相鄰的，也就是説是被隔離開來的，因此選擇B。

7. 答案：C。

本題考查原因解釋的題型中的解釋矛盾，題幹中存在的矛盾是「君主的長子擁有絕對的王位繼承優先權」與「現任女王在內，英國歷史上有很多女王都繼承了王位。」那麼A、B、D三個選項作為聽眾的誤解都可以成為這對矛盾的原因，而C選項無法成為這樣一種矛盾的原因，因此選擇C。

8. 答案：C。

第一步：抓住題幹的主要信息

題幹主要信息為食鹽來説，溫度升高了，溶解度並沒有多大的變化。

第二步：分析題幹信息，並結合選項得出答案。

由題幹主要信息可知，溫度升高對食鹽的溶解度沒有多大影響，C項用熱水溶解食鹽並非高人一等，推斷正確。A、D項錯誤；B項題幹未涉及到，故正確答案為C。

9.答案：C。

根據題幹給出的條件可以判斷丙所說「甲一定能通過」與戊所說「甲沒有可能通過」是矛盾關係，滿足「一真一假」，題幹中指出四句話中兩個人是真話，可以推出甲與丁所說也滿足一真一假，丁所說「乙能通過」如果為真，那麼甲所說也為真，與題幹所推出的結論相矛盾，由此可以確定丁所說一定為假，即乙不能通過，那甲所說為真，即甲與乙至少有一個能通過，已知乙不能通過，那甲就必然能通過考試，故推出C項。

10. 答案：A。

本題可以採用「排除法」。題幹要求每區種3種果樹，至少有一個區要同時種植杏和蘋果，桃樹和柿子樹也要同時種植。將選項帶入到題幹之後發現，只有A是錯誤的。A選項兩個區都種桃樹，因此也要種柿樹。題幹要求每區種植3種果樹，現在已經種植了桃樹和柿樹，因此只能選杏樹和蘋果樹的一種進行種植。如果滿足題幹第二個條件要求，則這個區就種植了4種果樹，與題幹第一個條件矛盾，因此A項是正確答案。

11. 答案：A。

第一步：找到論點和論據

本題論點是註冊結婚有避免兩人感情關係破裂後，導致的很多無謂糾紛的作用，論據是「未註冊結婚就共同生活在一起，在發生財產等民事糾紛時，不能得到法律上的承認和保護。」

第二步：判斷加強方式

本題論點和論據間存在因果關係，所以需加強論點。

第三步：逐一判斷選項的作用

A項：選項中「離婚率越來越高」說明男女感情不穩定，容易發生糾紛，最有力地加強了題幹觀點。

B、C項：說法均是對題幹觀點的削弱

D項：有些非婚同居者後來也去領了結婚證，在一定程度上加強了題幹觀點，但力度很弱。

綜上，A項的內容力度最強，故正確答案為A。

12. 答案：A。

由條件(1)知，有1個或者0人是罪犯。結合條件(2)如果甲是罪犯，乙肯定是共犯，可知甲不是罪犯。由條件(3)知道，丙不是罪犯。故甲、丙都不是罪犯。但不能推出乙一定是罪犯，可能存在沒有人是罪犯的情況，故正確答案為A。

13. 答案：C。
加強支持型題。題目中的結論是：收費方式的改變可以增加收入。按照之前的收費方式，如果停車不到兩小時，收費是2元，按照改變之後的收費方式，停車不到兩小時，收費是4元。如果C項為真，即很少有人會在停車場內一次停車超過兩小時，收入肯定是增加的。其他三項都對收入沒有影響，故正確答案為C。

14. 答案：D。
第一步：找突破口
「舉紅旗的人從西面上山」而「甲不舉紅旗」，所以甲不從西面上山，又因為甲「也不從東面上山」，所以甲只能從南面上山。因為「甲不舉紅旗」，「乙舉綠旗」，所以甲只能舉黃旗。
第二步：根據事實信息進一步推進
「舉紅旗的人從西面上山」而「乙舉著綠旗」，所以乙不從西面上山，只能從南面和東面上山，又因為從第一步可知甲從南面上山，所以乙只能舉綠旗從東面上山。
綜上所述，甲舉黃旗從南面上山，乙舉綠旗從東面上山，剩下丙舉紅旗從西面上山。故正確答案為D。

15. 答案：A。
第一步：從題幹中找論點和論據
論據：血液和體液中具有一種防凍蛋白，該蛋白使得它們得以存活並演化至今。
論點：該種魚類的生存卻面臨巨大挑戰是海水升溫導致的。
第二步：逐一分析選項
A項：在「從升溫對蛋白的影響」與「防凍蛋白對魚類生存影響」間搭建橋梁，能夠加強。
B項：「可有可無」是對論據的否定，不能支持。
C項：「魚類的多樣性」題幹未提及，是無關項。
D項：跟升溫無關，也是無關選項。
故正確答案為A。

16. 答案：C。
論據為每年都大量的人500萬死於煙草相關疾病。論點為煙草是首要的可預防死因。
A、B項：說明死於運動不足和生活不規律的人也很多，說明煙草並非首要可預防的，否定論點。

C項：死於並非煙草相關的疾病，可能存在許多其他的疾病，不能說明存在危害程度超過煙草的。

D項：直接說明心理調適不當是首要可預防死因，否定論點。題目中沒有加強的選項，因此選擇無關項最不能削弱。

17 答案：C。

題目中的結論是「本食品可放心食用」。理由是「不含防腐劑和人工色素」。要想結論正確，必須保證理由能推出結論，即：不含防腐劑和人工色素，本食品可放心食用，故正確答案為C。

18. 答案：C。

第一步：找到論點和論據

本題論點是日益改善的工作環境，可以提高工人的生產效率。論據是那些在舒適環境裡工作的人，比在不舒適環境裡工作的人生產效率高25%。

第二步：判斷削弱方式

本題論點和論據間存在因果關係，所以需削弱論點。

第三步：逐一判斷選項的作用

A項：只提到生產效率，沒有提到它與工作環境的關係，屬無關選項；B項：舒適的環境使員工努力工作，但努力工作不代表工作效率高，屬無關選項。

C項：說的是因為生產率高，才給予其舒適的工作環境，削弱了論點中的因果關係，是因果倒置。

D項：說的是生產效率高低與工作時間的關係，沒有提到工作環境，屬無關選項。故正確答案為C。

19. 答案：B。

本題屬於削弱題型：

第一步：找出論點論據

論點：要想以後的婚姻更穩固，必須延長戀愛時間。

論據：調查發現，戀愛時間越長的夫妻，婚後鬧分手的幾率會比那些戀愛時間較短的夫妻低。

第二步：逐一分析選項

B項：否定論點中的前提，說明戀愛時間長短跟婚姻牢固與否的關係不大，削弱了論點。

A項：結婚的可能性低不能得出婚姻不牢固的結論，無法削弱論點。
C項：速成婚姻容易破碎，說明婚姻牢固與戀愛時間長短有關，屬於加強選項。
D項：跟婚姻是否牢固無關，屬於無關選項。
故正確答案為B。

20. 答案：D
第一步：觀察題幹
如題目的題幹中涉及多個對象，題幹條件之間的邏輯關係不是很強時，最佳的答題方法是結合選項進行排除。
第二步：逐一排除選項
由「王、陳兩人至少有一人參加」可知，A項「王、陳不參加」不正確。
由「王、林、胡3人中應有兩人參加」可知，B項「林、胡不參加」不正確。
由「李、周兩人中也只要一人參加」可知，C項「周、李不參加」不正確。
故正確答案為D。

21. 答案：A。
第一步：找出論點和論據
論點：獎學金對幫助學生提高學習效率的作用是很明顯的
論據：獲得獎學金的學生，比那些沒有獲得獎學金的學生的學習效率要高出25%。
第二步：逐一判斷選項
A項：說明是因為學習習慣良好和學習效率高才獲得獎學金，直接削弱論點。
B項：認為獎學金能幫助學生改善學習環境提高學習效率，支持了論點。
C項：說學習效率低與缺少休息有關，但題幹沒涉及休息時間與學校效率的關係，屬無關選項。
D項：只是提出另一種研究方法，沒有說明具體問題，削弱不了結論。
故正確答案為A。

22. 答案：B。
第一步：抓住題幹主要信息
第一句介紹海豚是體型較小的鯨類，共有62種；第二句介紹海豚是一種本領超群、聰明伶俐的海中哺乳動物。
第二步：逐一分析選項
由第二句海豚是一種聰明伶俐的海中哺乳動物可推出，海洋中有哺乳動物是聰明伶俐的，即有些聰明伶俐的哺乳動物生活在海洋中，B項正確。

由題幹第一句可知，A項錯誤，海豚是體型較小的鯨類，而非體型較小的哺乳動物；第一句只説明海豚有62種，並沒有説明海洋中的哺乳動物的種類，C項無法得出；題幹沒有提及海豚是否善解人意，D項無法得出，故正確答案為B。

23. 答案：D。
第一步：抓住題幹的主要信息
題幹指出「養寵物的人往往身體比較健康，心理比較樂觀」。
第二步：逐一判斷選項的作用
A項：説「養寵物會讓人身心健康」，也就是説養寵物和身心健康間有因果聯繫，而題幹中並沒有體現這種因果聯繫，A錯誤。
B項：「身心健康的人才有精力養寵物」也就是説有精力養寵物是一定身心健康的人，而題幹説「養寵物的人往往身體比較健康」，B過於絕對，B錯誤。
C項：養寵物並不一定意味著「對動物有愛心」，C偷換概念，C錯誤。D項：由題幹「養寵物的人往往身體比較健康，心理比較樂觀」可知養寵物的人往往身心比較健康，D正確，故正確答案為D。

24. 答案：B。
本題為關聯詞推導中的推出方式類似題型，需要逐一分析題幹和選項的結構。
第一步：翻譯題幹
題幹：條件(1)所有睡蓮都是晚上開的，條件(2)這株蓮花是不是晚上開的，結論是這株蓮花不是睡蓮。
A項：條件(1)所有職業短跑運動員都穿釘鞋，條件(2)李先生不是職業運動員，結論是李先生不穿釘鞋。
B項：條件(1)所有山羊都長有鬍鬚，條件(2)這隻羊沒有鬍鬚，結論是這不是山羊。
C項：條件(1)所有醫生都在醫院工作，條件(2)王先生在醫院工作，結論是小王是醫生。
D項：條件(1)所有納税人都有存款，條件(2)這位太太有存款，結論是這位姑娘是納税人。
經比較與題幹結構一樣的是B項。故正確答案為B。

25. 答案：D。
第一步：找到論點論據
論點：非洲南方古猿像現代人一樣用雙腳行走
論據：非洲南方古猿具有定型的弓形足

第二步：判斷加強方式

論點強調南方古猿像現代人一樣用雙腳行走，論據強調南方古猿具有定型的弓形足，兩者範圍不同，沒有明顯的因果關係，需要在兩者間搭橋，加強論證。

第三步：逐一分析選項

D項：從具有定型的弓形足到像現代人一樣用雙腳行走之間建立聯繫，加強了論證，是答案。

A項：說明得到論據的方式，加強了論據，但不能說明論點和論據之間的關係，無法加強論點。

B項：是否只有南方古猿是人類的祖先，跟南方古猿是否像現代人一樣用雙腳行走無關，無法加強論點。

C項搭橋方向相反。

故正確答案為D。

26. 答案：C。

第一步：找到論點和論據

本題論點是鎮靜劑對降低日常生活中的心理壓力也會很有效，論據是由測謊器測出的心理壓力反應可以通過藥物抑制，而不會產生顯著的副作用。

第二步：判斷加強方式

論點強調日常生活中的心理壓力，論據強調測謊器測出的心理壓力，所以論點與論據間不存在明顯因果關係，所以需加強論證，在論點和論據間搭橋。

第三步：逐一判斷選項的作用

C即在測謊器測出的心理壓力與日常生活的心理壓力間搭橋，是答案。A、B、D均與日常生活中的心理壓力無直接關係，屬無關選項，故正確答案為C。

27. 答案：B。

第一步：抓住題幹主要信息

第一句說明十六世紀英國貨幣市場上出現的不足值貨幣驅逐足值貨幣的現象。第二句說明這是一種「優敗劣勝」的自然與社會現象。

第二步：根據題幹信息，逐一分析選項

B項：山寨手機取代正品手機流行於市場上，屬於題幹中提到的「優敗劣勝」的現象。

A項：猴群中強者為王

C項：優秀作品盛傳

D項：品學兼優的學生突出，體現的是優勝劣敗的現象，與題幹相反。故正確答案為B。

28. 答案：D。

題幹的論證是「受檢水獺都與人及其寵物有接觸，其患病率為2%→該省水獺患病率大大小於2%」。如果D項的斷定為真，根據題幹「只有與人及其寵物有接觸的水獺才接受檢疫」則説明在接受檢疫的水獺中感染大瘋狗症的比例，要高於未接受檢疫的水獺。這就有力地支持了專家的推測。其他各項均不能支持專家的推測，故本題選D。

29. 答案：D。

觀察到題幹問「最可能是這幾個廠的產品選擇方案」可知，本題正面分析不易，因此可採用代入選項排除的方法解決。

甲廠不擅長生產消毒櫃，因此排除A項。

乙廠生產冷氣機和消毒櫃，與丙廠衝突，因此排除B項。

丙廠不擅長生產微波爐，因此排除C項。

D項最為合理，乙廠只生產洗衣機，丙廠生產空調和消毒櫃，剩下的甲廠可生產電冰箱和微波爐，均是各自所擅長的。

綜上，故正確答案為D。

30. 答案：C。

A、B項屬於支持項，排除。D項屬於無關項，不能削弱結論。C項指出流感致病細菌的傳播與試驗對像大腸杆菌不同，直接否定了專家的推論。

PART

TWO

Verbal Reasoning

關於 Verbal Reasoning

Verbal Reasoning(語文推理) 是根據一篇短文，來判斷題幹信息正確與否，主要考察應聘者的英語閱讀能力和邏輯判斷能力。

Verbal Reasoning的選項有三：

一是Yes，就是説題幹的信息根據原文來判斷是正確的；

二是No，就是説題幹的信息根據原文來判斷是錯誤的；

三是Can’t tell，就是根據原文提供的信息無法判斷對錯。

答題技巧

1. 解題步驟

Step 1：定位，找出題目在原文中的出處。

(1)找出題目中的關鍵詞， 最好先定位到原文中的一個段落。

(2)從頭到尾快速閱讀該段落，根據題目中的其他關鍵詞，在原文中找出與題目相關的一句或幾句話。

(3)仔細閱讀這一句話或幾句話，根據第二大步中的原則和規律，確定正確答案。

(4)要注意順序性，即題目的順序和原文的順序基本一致。

Step 2：判斷，根據以下原則和規律，找出正確答案。

2. True的特點

(1)題目是原文的同義表達。通常用同義詞或同義結構。

例題1：

原文：Few are more than five years old. (很少有超過五年的)

題目：Most are less than five years old. (大多數都少於五年)

答案：題目與原文是同義結構，所以答案應為True。

例題2：

原文：Frogs are losing the ecological battle for survival, and biologists are at a loss to explain their demise. (青蛙失去了生存下來的生態競爭能力，生物學家不能解釋它們的死亡。)

題目：Biologists are unable to explain why frogs are dying. (生物學家不能解釋為什麼青蛙死亡)

答案：題目中的are unable to與原文中的are at a loss to是同義詞，題目中的why frogs are dying與原文中的their demise是同義詞，所以答案應為True。

(2)題目是根據原文中的幾句話做出推斷或歸納。不推斷不行，但有時有些考生會走入另一個極端，即自行推理或過度推理。

例題1：

原文：Compare our admission inclusive fare and see how much you save. Cheapest is not the best and value for money is guaranteed. If you compare our bargain Daybreak fares, beware - most of our competitors do not offer an all inclusive fare. (比較我們包含的費用會看到你省了很多錢。最便宜的不是最好的。如果你比較我們的價格，會發現絕大多數的競爭對手不提供一籃子費用。)

題目：Daybreak fares are more expensive than most of their competitoRs.(Daybreak的費用，比絕大多數的競爭對手都昂貴。)

答案：雖然文章沒有直接提到的費用比絕大多數的競爭對手都昂貴。但從原文幾句話中可以推斷出Daybreak和絕大多數的競爭對手相比，收費更高，但服務的項目要更全。與題目的意思一致，所以答案應為True。

例題2：

原文：For example, it has been demonstrated that rapid response leads to a greater likelihood of arrest only if responses are in the order of 1-2 minutes after a call is received by the police. When response times increase to 3-4 minutes - still quite a rapid response - the likelihood of an arrest is substantially reduced. (例如，只有反應時間在警察接到電話之後1至2分鐘，快速反應才會使抓住罪犯的可能性更大。當反應時間增加到3至4分鐘，仍然是非常快的反應，抓住罪犯的可能性就實質性的降低。)

題目：A response delay of 1-2 minutes may have substantial influence on whether or not a suspected criminal is caught. (1至2分鐘的反應延遲會對嫌疑犯是否被抓住產生實質性的影響。)

答案：從原文的兩句話可以推斷出：1至2分鐘，抓住罪犯的可能性很大，3至4分鐘，可能性就實質性的降低。所以，1至2分鐘的反應延遲會對嫌疑犯是否被抓住產生實質性的影響，答案應為True。

3. False的特點

(1)題目與原文直接相反。通常用反義詞、not加同義詞及反義結構。no longer / not any more / not / by no means……對比used to do sth. / until recently /as was once the case。

例題1：

原文：A species becomes extinct when the last individual dies. (當最後一個個體死亡時，一個物種就滅亡了。)

題目：A species is said to be extinct when only one individual exists. (當只有一個個體存活時，一個物種就被說是滅亡了。)

答案：可以看出題目與原文是反義結構。原文説一個物種死光才叫「滅絕」，而題目説還有一個個體存活就叫「滅絕」，題目與原文直接相反，所以答案應為False。

例題2：

原文：It has been successfully used in the United States to provide input into resource exploitation decisions and assist wildlife managers and there is now enormous potential for using population viability to assist wildlife management in Australia’s forests.(在美國它已經成功地用於支持資源開發和幫助野生生命研究管理者。現在，在使用它對澳洲的森林中的野生生物管理上，有巨大的潛力。)

題目：PVA has been used in Australia for many yeaRs.(PVA已經在澳洲使用多年了)

答案：原文說PVA在澳洲的研究中有巨大的潛力，即剛剛開始。題目說在澳洲已經使用多年，所以題目與原文是反義結構，答案應為False。

(2)原文是多個條件並列，題目是其中一個條件(出現must或only)，原文是兩個或多個情形(通常是兩種情形)都可以，常有both…and、and、or及also等詞。以及various / varied / variety / different / diversified / versatile等表示多樣性的詞彙。題目是「必須」或「只有」或是「單一」其中一個情況，常有must及only / sole / one / single等詞。

例題1：

原文：Booking in advance is strongly recommended as all Daybreak tours are subject to demand. Subject to availability, stand by tickets can be purchased from the driver. (提前預定是強烈建議的，因為所有的Daybreak旅行都是由需求決定的。如果還有票的話，可直接向司機購買。)

題目：Tickets must be bought in advance from an authorized Daybreak agent. (車票必須提前從認可的代理購買)

答案：原文是提前預定、直接向司機購買都可以，是多個條件的並列。題目是必須提前預定，是必須其中一個情況。所以答案應為False。

例題2：

原文：Since the Winter Games began, 55 out of 56 gold medals in the men’s Nordic skiing events have been won by competitors from Scandinavia or the former Soviet Union. (自從冬季奧運會開始，在男子越野滑雪項目中的56塊中的55塊金牌被，被來自北歐和前蘇聯的選手奪得。)

題目：Only Scandinavians have won gold medals in the men’s

winter Olympics. (只有北歐人獲得了冬運會男子越野滑雪項目中的金牌。)

答案：原文是北歐人和前蘇聯的選手獲得了金牌，而且是獲得了56中的55塊，還有1塊不知道被誰獲得。題目是只有北歐人獲得了金牌。所以答案應為False。

(3)原文強調是一種「理論」(theory)、「感覺」(felt)、「傾向性」(trend / look at the possibilities of)、「期望或是預測」(it is predicted / expected / anticipated that)等詞。而題目強調是一種「事實」，常會用到real / truth / fact / prove等字詞。

例題1：

原文：But generally winter sports were felt to be too specialized. (但一般來説，冬季項目被感覺是很專門化的。)

題目：The Antwerp Games proved that winter sports were too specialized. (Antwerp運動會證明冬季項目是很專門化的)

答案：原文中有feel，強調是「感覺」。題目中有prove，強調是「事實」，故答案應為False。

例題2：

原文：Another theory is that worldwide temperature increases are upsetting the breeding cycles of frogs. (另一種理論是世界範圍溫度的升高破壞了青蛙的生長循環。)

題目：It is a fact that frogs' breeding cycles are upset by worldwide increases in temperature. (一個事實是青蛙的生長循環被世界範圍溫度的升高所破壞。)

答案：原文中有theory，強調是「理論」。題目中有fact，強調是「事實」。所以答案應為False。

(4)原文和題目中使用了表示不同程度、範圍、頻率、可能性的詞。原文中常用typical、odds、many、sometimes及unlikely等詞。題目中常用special、impossible、all、usually、always及impossible等詞。

例題1：

原文：Frogs are sometimes poisonous. (青蛙有時是有毒的)

題目：Frogs are usually poisonous. (青蛙通常是有毒的)

答案：原文中有sometimes，強調是「有時」。題目中有usually，強調是「通常」，所以答案應為False。

例題2：

原文：Without a qualification from a reputable school or university, it is unlikely to find a good job. (不是畢業於著名學校的人，都不太可能找到一個好的工作。)

題目：It is impossible to get a good job without a qualification from a respected institution. (不是畢業於著名學校的人，找到一個好的工作是完全不可能的。)

答案：原文中有unlikely，強調是「不太可能」。題目中有impossible，強調是「完全不可能」。所以答案應為False。

(5)情況原文中包含條件狀語，題目中去掉條件成份。原文中包含條件狀語，如if、unless或if not也可能是用介詞短語表示條件狀語如in、with或but for。題目中去掉了這些表示條件狀語的成份。這時，答案應為False。

例題：

原文：The Internet has often been criticized by the media as a hazardous tool in the hands of young computer useRs.(互聯網被媒體評為年輕電腦用戶的危險工具。)

題目：The media has often criticized the Internet because it is dangerous. (媒體經常指責互聯網，因為它是危險的。)

答案：原文中有表示條件狀語的介詞短語in the hands of young computer users，題目將其去掉了。所以答案應為False。

(6)出現以下詞彙，題目中卻沒有說明：less obviously/ less likely/ less possible。

4. Can't tell的特點

(1)題目中的某些內容在原文中並沒有提及，而題目中的某些內容在原文中亦找不到依據。

(2)題目中涉及的範圍小於原文涉及的範圍，也就是更具體。原文涉及一個較大範圍的範疇，而題目是一個具體概念。也就是説，題目中涉及的範圍比原文要小。

例題1：

原文：Our computer club provides printer. (我們電腦學會有提供打印機的服務)

題目：Our computer club provides color printer. (我們電腦學會有提供彩色打印機的服務)

答案：題目中涉及的概念「比原文中涉及的概念」要小。換句話説，電腦學會有提供打印機的服務，但至於是彩色還是黑白，不知道或有可能，文章中沒有給出進一步的信息。所以答案應為Can' t tell。

例題2：

原文：Tourists in Cyprus come mainly from Asia. (前往塞浦路斯旅遊的遊客，主要來自亞洲。)

題目：Tourists in Cyprus come mainly from the Korea. (前往塞浦路斯旅遊的遊客，主要來韓國。)

答案：題目中涉及的概念「Korea」比原文中涉及的概念「Asia」要小。原文只說到塞浦路斯旅行的遊客主要來自亞洲，有可能主要來自韓國，但亦有可能主要來自亞洲內其他國家，文章中沒給出進一步的信息。所以答案應為Can't tell。

(3)原文是某人的目標、目的、想法、願望、保證、發誓等，題目是事實。原文中常用aim / goal / promise / swear / vow / pledge / oath / resolve等詞。題目中用實意動詞。

例題1：

原文：He vowed he would never come back. (他發誓將永不回來)

題目：He never came back. (他沒再回來)

答案：原文中說他發誓將永不回來，但實際怎麼樣，不知道。也可能他違背了自己的誓言，所以答案應為Can't tell。

例題2：

原文：His aim was to bring together, once every four years, athletes from all countries on the friendly fields of amateur sport. (他的目的是把各國的運動員，每四年一次聚集到友好的業餘運動的賽場上。)

題目：Only amateur athletes are allowed to compete in the modern Olympics. (只有業餘運動員被允許在現代奧運會中參賽。)

答案：原文中用aim表示「目的」，題目中用實意動詞表示「事實」。把各國的運動員聚集到友好的業餘運動的賽場上，這只是創建者的目的，實際情況如何，文章中沒說，所以答案應為Can't tell。

(4)原文中沒有比較級，題目中有比較級。

例題：

原文：In Sydney, a vast array of ethnic and local restaurants can be found to suit all palates and pockets. (在悉尼，有各種各樣的餐廳。)

題目：There is now a greater variety of restaurants to choose from in Sydney than in the past. (現時有更多種類的餐廳，供食客選擇。)

答案：原文中提到了悉尼有各種各樣的餐廳，但並沒有與過去相比，所以答案應為Can’t tell。

(5)原文中是虛擬would / even if，題目中卻是事實。（虛擬語氣看到當作沒有看到）

(6)原文中是具體的數據事例，而題目中卻把它擴大化，規律化。

5. 其他實用技巧

(1)原文是判斷的唯一根據

原文是判斷的唯一根據。所以，無論你對文章的內容或背景多麼熟悉，或者你的知識多麼豐富，都不能憑借自己的知識來確定答案。即使題目中說「地球是正方體的」。如果原文中沒有提及，你也只能選Can't tell，不能選False。

(2)不能做無根據的自行推斷

有些題目需要根據原文做適當的判斷和推理，才能確定正確的答案，但必須根據原文來做推斷，而不能做毫無根據的推理。而且一般來講，即使需要推理，也只需要推理一步即可，在這類題目中，推理太深反而容易做錯。

(3)注意題目要求答什麼

有些考生做題目的時候，尤其是規定的時間比較短的情況下，容易下意識地忽略題目而直接進行選擇，這樣容易導致判斷正確卻失分的情況，非常可惜。避免這種情況的方法是：平時練習中就有意識地訓練自己按照要求答題的習慣。

(4)題目中若出現must、only、all、always，答案一般不會是True

題目中會經常出現這類的詞，而出現這類詞的題目，其答案95%都不是True。其答案是False還是Can’t tell要視乎情況而定，但若不是自己已非常確定是True的情況下，請不要選擇True，一般來說是False的情況居多。

(5)相信自己的第一感覺

在考試中，除非有特別充分的理由，否則不要輕易更改答案，人的第一直覺往往是正確的。而很多同學都將正確的答案改成了錯誤的。

以上幾點方法和規律的運用不能鑽牛角尖。這種題型本身就有一定的缺陷。所以上述的規律和方法如果不能理解的話，就記住它們。考試的時候，按照這些規律來做題目就可以了，因為這些規律都經過大量實踐的檢驗，符合大多數的情況。

試題練習

Passage 1

The project was ambitious in its size, complexity, triparty nature, and in its pioneering of the Private Finance Initiative. This difficulty was unavoidable and contributed to the project's failure. However, a more thorough estimate of the unknown difficulties and timescales would have enabled the Department to better prepare for the project, and increase its chance of success.

In December 1997 XSoft indicated they needed time to complete the project, which should have been inevitable. If the Department knew from the start how long the project would take, it is questionable whether they would have considered inception, especially considering the implications of delay on the overall profitability for the venture.

1. If more care had been put into estimating the difficulties, it is less likely the project would have failed.

2. XSoft witheld information from the Department regarding how long the project would take.

3. The Department's profits were dependent upon how long the project took.

Passage 2

Ever since the gun's invention it has been changing the world in many different ways. Many of the developments in gun design have been brought about by man's desire to protect himself, and the challenge of inventing bigger and more accurate weapons.

Each time there has been a major innovation in the development of the gun, there has been a profound effect on the world. The gun helped in the exploration of the world, it has also helped in the development of society as we know it.

4. The gun was invented because the human race needs to protect themselves.

5. Guns are the reason our society is the way it is today.

6. Financial incentives had no part to play in the development of the gun.

Passage 3

Being socially responsible is acting ethically and showing integrity. It directly affects our quality of life through such issues as human rights, working conditions, the environment, and corruption. It has traditionally been the sole responsibility of governments to police unethical behavior. However, the public have realized the influence of corporations and, over the last ten years, the level of voluntary corporate social responsibility initiatives that dictate the actions of corporations has increased.

7. The ethical actions of corporations has changed over the last ten years.

8. Corporations can influence the public's quality of life.

9. Traditionally, the government have relied upon only the large corporations to help drive corporate social responsibility, whilst they concentrated on the smaller corporations.

Passage 4

A well-nourished child can be more likely to be a studious one. But food has been seen as a cost to be cut in times of austerity, rather than an ingredient of good schooling. That may now be changing: as the government worries about obesity – which is fast rising among children- and urges everyone to eat less salt, fat and sugar, and more fruit and vegetables, the deficiency and unhealthiness of most school meals is striking. But cash constraints make change difficult.

10. Children who eat healthily will perform better in exams.

11. The number of obese children used to be less than it is now.

12. The government is apathetic about obesity.

Passage 5

The costs of roaming - the service which allows UK customers to use their mobile phone abroad - are much higher than those in France, Germany, Sweden and Italy. Many people get caught out because they are unaware of the high prices, and that they get charged for simply receiving calls whilst abroad. Less than a quarter of consumers had any knowledge of the price of using a mobile phone aboard when they bought their phone. Better consumer information is vital if prices for pre-pay international roaming in the UK are to come down.

13. It is more expensive for German customers to use a roaming service than it is for UK customers.

14. Pay-as-you-go roaming rates are lower than they are for contract customers.

15. Customers can be charged for calls which they do not make themselves.

Passage 6

Globalization is putting fresh pressure on mid-sized firms, with large companies moving into their markets and dictating prices. This kind of company feels the pressure on all sides. They are too large to qualify for the grants and assistance that small enterprises can count on, but are too small to wield the kind of influence their larger competitors can bring to bear. But despite the squeeze, most mid-sized firms still believe that they can deliver steady growth, by playing to their traditional strength: being more nimble and more customer-focused than their larger rivals.

16. Large companies have a greater effect on prices than mid-sized companies.

17. The size of grant awarded to a company is indirectly proportional to the size of the company.

18. Traditionally mid-sized firms are faster than large companies at adapting to market changes.

Passage 7

If society seems obsessed with youth, it is at least partly because companies are. Like it or not, the young increasingly pick the styles and brands that trickle up to the rest of the population. Nike, Abercrombie & Fitch and Timberland first found success with the young, and when that clientele tired of them the companies felt the loss deeply. Now that adults are no longer necessarily expected to act and look grown-up, parents and children can be found listening to exactly the same music, playing the same computer games, watching the same TV programs, and wearing the same brands of clothes and shoes.

19. An adult's style can sometimes be similar to that of a child's.

20. The profits of Timberland are not affected by young customers.

21. Adults wear the same shoes as children because they want to look younger.

Passage 8

Television is changing as it goes digital. The result will not only be better-quality pictures and sound but also personal TV, with viewers able to tailor the programs they watch and even interact with them. How much money this will make for program producers or broadcasters, whoever they may be, is not so clear.

Cable, satellite and terrestrial television broadcasters are upgrading their equipment to provide higher quality digital services. Rupert Murdoch's News Corporation will become the first company in the world to migrate an entire national TV system over to digital when it turns off its old analogue version of its British satellite service, BSkyB.

22. Rupert Murdoch is associated with BSkyB.

23. The only change from traditional analogue services to digital services will be the picture quality.

24. Television broadcasters are upgrading their equipment because they will make more money from digital TV.

Passage 9

Brand equity has become a key asset in the world of competitive business. Indeed, some brands are now worth more than companies. Large corporations themselves are widely distrusted, whereas strangely, brands have the opposite effect on people. Brands are used to humanize corporations by appropriating characteristics such as courage, honesty, friendliness and fun. An example is Dove soap, where a dove represents white, cleanliness and peace. Volkswagen like to give the impression through their advertising that they are a reliable, clever, technical product. In a sense, rather than the product itself, the image

25. Brands have always been an important asset to a company.

26. Many people distrust large corporations.

27. Dove soap chose a dove for their brand to give a sense of cleanliness and peace.

Passage 10

The first problem with financial statements is that they are in the past; however detailed, they provide just a snap-shot of the business at one moment in time. There is also a lack of detail in financial statements, giving little use in the running of a business. Financial statements are provided for legal reasons to meet with accounting regulations and are used mainly by City analysts who compute share prices and give guidance to shareholdeRs.Accounts often have hidden information and may also be inconsistent; it is difficult to compare different companies' accounts, despite there being standards, as there is much leeway in the standards.

28. Financial statements are useful for businesses to understand their financial activities.

29. Companies create financial statements in order to comply with their legal obligations.

30. If account reporting standards were tightened, it would be easier to compare the performance of different companies.

答案及解析

1. Answer: True.

Since more thorough can be considered equivalent to giving more care, this is stated as true in the following excerpt "a more thorough estimation of the unknown difficulties and timescales would…increase its chance of success".

2. Answer: Can't tell.

We are told XSoft requested more time, and the passage implies the Department did not know how long it would take at the beginning, but the passage does not tell us if XSoft did or did not withhold time information.

3. Answer: True

We are told that there were "implications of delay on overall profitability for the venture".

4. Answer: Can't tell

The passage does not say how or why the gun was invented. It does say that some changes to the gun's design have been because humans want to protect themselves, but the passage does not say how or why the gun was first invented.

5. Answer: False

The gun "has also helped in the development of society as we know it". The word help implies it is not the only contributor and is therefore not the reason our society is the way it is today.

6. Answer: Can't tell

The passage does not mention financial incentives or economic gain, so we cannot tell from information in the passage alone.

7. Answer: True

The passage says that the public have caused corporations to alter their ethical actions: "over the last ten years, the level of voluntary corporate social responsibility initiatives that dictate the actions of corporations has increased."

8. Answer: True

This one is slightly less obvious. We are first told that being socially responsible directly affects our quality of life. Then we are told that corporate social responsibility dictates the actions of corporations. So following that logic, corporations must be able to affect our quality of life.

9. Answer: False

We are told that "traditionally it has been the sole responsibility of the government to police unethical behavior", meaning traditionally no corporation played a part; not even large corporations.

10. Answer: Can't tell

The passages says "a well-nourished child can be more likely to be a studios one", which means they are more likely to study. Whilst the passage suggests eating healthily can help, it does not say definitively if the child will or will not perform better in exams.

11. Answer: True

The passage says, in reference to obesity, "is fast rising among children" meaning in the past it was less than it is now.

12. Answer: False

The passage says "...as government worries about obesity...". Worrying about something is the opposite of being apathetic about something.

13. Answer: False

The opposite is true since the passage says "the costs of [UK] roaming...are much higher than those in France, Germany"

14. Answer: Can't tell

The passage does not tell us anything about the price of roaming for contract phone customers, so we cannot compare these prices to pre-pay or pay as you go roaming charges. Without further information we cannot say for sure.

15. Answer: True

The passage states that customers "get charged for simply receiving calls while abroad".

16. Answer: True

The first sentence says that large companies dictate the prices for mid-sized firms.

17. Answer: Can't tell

With the pressure of time, it might be tempting to answer True here. On careful reading of the passage, it describes how mid sized firms are too large to qualify for grants, but it does not say explicitly how the size of a grant might be related to the size of company. The question is stating quite a specific relationship, whereas all we are told is that small companies get grants and mid-sized firms do not. For example we are told nothing about the size of a grant a large company might get.

18. Answer: True

The last sentence describes one of the traditional strengths of a mid-size firm as "being more nimble" which we can interpret to mean faster at adapting to change.

19. Answer: True

The sentence that proves this is "parents and children can be found listening to exactly the same music, playing the same videogames...and wearing the same brands of clothes and shoes [as the young]".

20. Answer: False

The passage says that when "when that clientele [the young] tired of them the companies felt the loss deeply". Meaning that Timberland's profits are in fact closely affected by the purchasing habits of the young.

21. Answer: Can't tell

The passage does not say why adults and children sometimes end up with the same styles. The passage implies that by listening to the same music and wearing the same clothes as young people, adults are not considered grown-up. But the passage does not say why some adults adopt this style (for example it could be price, comfort, or any number of reasons other than wanting to look younger, but we are not told.)

22. Answer: True
This is stated it he passage where it says "Rupert Murdoch's News Corporation... BSkyB", meaning he is therefore associated.

23. Answer: False
The passage says of switching from analogue to digital: "The result will be better quality pictures and sound but also personal TV, with viewers able to tailor the programs they watch...".

24. Answer: Can't tell
This is quite obvious since the passage explicitly says in the last sentence of the first paragraph that it is not clear how much money broadcasters will make from upgrading.

25. Answer: Can't tell
The passage says "Brand equity has become a key asset in..."
implying it has not always been. However "important" is distinct from key, and the passage does not discuss if brand equity was considered important before being considered key.

26. Answer: True
The passage says "Large corporations themselves are widely distrusted". Widely implies an opinion held by many people, and the opinion held would be one of distrust.

27. Answer: Can't tell
Whilst the passage does say that the dove gives an impression of "white, cleanliness and peace", and therefore is a successful brand, the passage does not say how or why the company chose the brand in the first place.

28. Answer: False
The passage says that financial statements have "a lack of detail in financial statements, giving little use in the running of a business", which is supported by other

critical statements such as 'the first problem with financial statements is that they are in the past" means they cannot be considered useful for businesses to understand their financial activities.

29. Answer: True

This is stated here: "Financial statements are provided for legal reasons to meet with accounting regulations".

30. Answer: Can't tell

The last sentence in the passage says, "it is difficult to compare different companies' accounts, despite there being standards, as there is much leeway in the standards". Though this implies that if there were less leeway in the standards, it would be easier to compare different companies' accounts, this is only an inference. Thus, it can only be asserted as a probability rather than a certainty. Since the passage does not expressly say, we require more information. Therefore, it is cannot say.

PART

THREE

Data Sufficiency Test

答題技巧

1. 背出答案選項

做過Data Sufficiency Test(數據充分題，簡稱「DS Test」)的考生都知道，DS Test的答案選項通常固定在以下5個：

A. 條件(1)單獨充分，條件(2)單獨不充分。

B. 條件(2)單獨充分，條件(1)單獨不充分。

C. 條件(1)和條件(2)一起充分，單獨都不充分。

D. 條件(1)和條件(2)分別單獨充分。

E. 條件(1)和條件(2)一起都不充分。

由於答案固定，建議考生完全可以把5個選項都背出來，而不需要在考試時再浪費時間去看一遍，這樣無形中會節省出更多答題時間。

2. 使用排除法

承上，即使能夠在5個答案中排除一個，考生仍然需要在剩下的4個選項中進行選擇，排除法本身的意義其實並不算太大。而DS Test則完全不同，排除一個答案往往意味著同時排除了數個關聯答案，能夠大大減少可選選項，提高選擇的命中率。具體來説：

情況1：條件(1)單獨充分，條件(2)未知。可排除B、C、E，可能選項只剩A或D。

情況2：條件(1)單獨不充分，條件(2)未知。可排除A、D，可能選項只剩B、C或E。

情況3：條件(1)未知，條件(2)單獨充分。可排除A、C、E 可能選項只剩B或D。

情況4：條件(1)未知，條件(2)單獨不充分。可排除B、D，可能選項只剩A、C或E。

以上幾種情況，無論哪種都意味著大量排除選項，瞬間縮小選擇範圍。因此，排除法在解答DS Test題目時，是十分高效實用的解題技巧。

3. 不做多餘計算

DS Test 於其他數學題目最大的不同之處就在於其不求甚解的特點。

DS Test 從不會要求考生做十分具體的計算，而常常需要一個判斷的結果或者大致範圍的了解，目的是在於確認條件單獨或組合是否成立。因此，考生在解題時一般都不需要做太多具體的計算工作。而只需要根據給出條件稍加判斷即可得出結論。如果你發現自己算了一大堆數據，那麼你的解題思路可能已經出現了問題。

4. 兩個條件單獨考慮

不少考生在做DS Test時，很容易犯的一個慣性錯誤就是直接把兩個條件放到一起考慮，或者在使用一個條件時無意識的同時代入另一個條件。因此，考生在面對此類題目時，一定要學會並習慣分開考慮條件的解題思路，把答題方式糾正過來，如果大家在做DS Test時也出現這種問題，那麼請務必通過平時練習養成正確的解題習慣。

以上這些技巧，如果能夠熟悉並掌握，對於大家在解答DS Test時，應該會有所幫助。希望各位考生都能認真參考學習本文內容，最終在DS Test中發揮出色。

下列的27條試題講解，均統一設有5個答案選項：

A. 條件(1)單獨充分，條件(2)單獨不充分。

B. 條件(2)單獨充分，條件(1)單獨不充分。

C. 條件(1)和條件(2)一起充分，單獨都不充分。

D. 條件(1)和條件(2)分別單獨充分。

E. 條件(1)和條件(2)一起都不充分。

試題講解

1. What percent of a group of people are women with red hair?

(1) Of the women in the group, 5 percent have red hair.

(2) Of the men in the group, 10 percent have red hair.

題目問：「紅頭髮女人在這群人當中佔的百分比？」

想要知道百分比，首先需要知道人群的總數，以及紅髮女人的數目。條件(1)說女人中有5%的女人為紅髮，條件(2)說有10%的男人是紅髮。兩個條件都得不出紅髮女人所佔的比例，故答案選E——兩個條件都不充分。

2. In a certain class, one student is to be selected at random to read. What is the probability that a boy will read?

(1) Two-thirds of the students in the class are boys.

(2) Ten of the students in the class are girls.

題目問：「課堂上隨機選擇一個學生閱讀，那麼選到男生閱讀的概率是多少？」

考生需要知道男生所佔全班人數的比率，才能知道男生閱讀的概率，故條件(1)充分，概率是三分之二。條件(2)只說明女生有10個，跟題幹的男生沒有任何關係，故答案選A。

3. If the two floors in a certain building are 9 feet apart, how many steps are there in a set of stairs that extends from the first floor to the second floor of the building?

(1) Each step is foot high.

(2) Each step is 1 foot wide.

題目問：「在一座建築物內，兩層樓梯間的距離9呎。那麼，從第一個樓梯到第二個樓梯，共有多少級台階？」

考生只需要知道每個台階的高度，就可以計算出台階的數目。條件(1)説明每級台階有0.75呎高，條件充分。條件(2)表示每級台階闊1呎，與題幹沒有相關性，故選A——只有條件(1)符合。

4. In a college, the number of students enrolled in both a chemistry course and a biology course is how much less than the number of students enrolled in neither?

(1) In College X there are 60 students enrolled in a chemistry course.

(2) In College X there are 85 students enrolled in a biology course.

題目問：「院校內既學習化學，又學習生物的學生數目，比兩個科目都不學的學生數量少幾多？」

考生需要知道兩樣都學的人數多少，兩樣都不學的人數又有多少。條件(1)指出上有60位化學堂的學生，條件(2)表示讀生物科的有85人。兩個條件不論怎麼配搭都求不出來，故答案是E——兩個條件都不符合。

5. If n is an integer, is n+1 odd?

(1) n+2 is an even integer

(2) n-1 is an odd integer

題目問：「如果n是整數，那麼n+1是奇數嗎？」

條件(1)說n+2是偶數，條件(2)說n-1是奇數。考生想要知道n+1是奇數，那麼滿足n是偶數就可以。條件(1)得出n是偶數，條件充分。條件(2)得出n是偶數條件也滿足，所以兩個條件都是單獨滿足，故答案選D。

6. A citrus fruit grower receives $15 for each crate of oranges shipped and $18 for each crate of apple shipped. How many crates of oranges did the grower ship last week?

(1) Last week the number of crates of oranges that the grower shipped was 20 more than twice the number of crates of apple shipped.

(2) Last week the grower received a total of $38,700 from the crates of oranges and apple shipped.

題目問：「種橙的果農每發一箱橙獲得$15，而每發一箱蘋果則可獲$18，請問這名果農上星期共發了多少箱橙？」

條件(1)：上周橙的數量，是蘋果的兩倍多20個。

條件(2)：上周總共收獲了38,700元

想要知道橙的數量，要是知道蘋果的數量就可以用條件(1)算出來，但我們不知道蘋果的數量，所以條件(1)單獨不能滿足。條件(2)給出總收獲是$38,700。要是知道橙買了多少錢的話，也可以求出數量。

假設橙是x，蘋果是y，那麼可以列出一條方程：15x+18y = 38700，X = 2y+20。此方程可解，所以可以求出有多少個橙，故答案選C——兩個條件結合才能得出答案。

7. Water is pumped into a partially filled tank at a constant rate through an inlet pipe. At the same time, water is pumped out of the tank at a constant rate through an outlet pipe. At what rate, in gallons per minute, is the amount of water in the tank increasing?

(1) The amount of water initially in the tank is 200 gallons.

(2) Water is pumped into the tank at a rate of 10 gallons per minute and out of the tank at a rate of 10 gallons every 2.5minutes.

題目問：「水流以恆定的速度進入蓄水池，並以恆定的速度流出蓄水池。那麼以多大的速度進入水池，才能保證水池的水是增加的？」

條件(1)：水池原始水量有200加侖

條件(2)：進水按照1分鐘10加侖的速度，出水按照2.5分鐘10加侖的速度。

只要保證進水速度大於出水速度，就可以滿足水池水量增加，所以可以看出條件(2)正確，答案選B。

8. What percent of the drama club members enrolled at a certain school are female students?

(1) Of the female students enrolled at the school, 40 percent are members of the drama club.

(2) Of thc malc students enrolled at the school, 25 percent are members of the drama club.

題目問：「某學校參加戲劇社的學生中，女學生佔多少？」

條件(1)：這個學校學生中40%都參加了戲劇社

條件(2)：學校男生中，有25%參加了戲劇社。

想要知道女生的比例，需要知道參加戲劇社學生總數是多少，男、女生是多少。條件(1)只表示參加戲劇社的學生有多少，卻不知道男女比例，條件(2)給出該校男生的25%參加戲劇社，還是不能得出答案，故答案是E。

9. How many people are directors of both Company K and Company R?

(1) There were 17 directors present at a joint meeting of the directors of Company K and Company R, and no directors were absent.

(2) Company K has 12 directors and Company R has 8 directors.

題目問：「有多少人既是X公司的總監，也是Y公司的總監？」

條件(1)：有17人參加了由K公司和R公司聯合組成的董事會，無人缺席。

條件(2)：K有12名董事會成員，R有8名。

條件(1)給出兩公司總共有17人，條件(2)給出董事會分別是多少，明顯多出來的3個人，就是兩間公司的共同董事會，故答案選C。

10. If the sequence S has 300 terms, what is the 293rd term of S?

(1) The 298th term of S is –616, and each term of S after the first is 2 less than the preceding term.

(2) The first term of S is –22.

題目問：「S序列有300個數，那麼第293個是多少？」

條件(1)：第298個是-616，並且S序列中每個數都比前一個數少2。

條件(2)：第一個數是-22

一個數列，只要知道數列之間數的關係，就可以按照規律求出每個數是多少。條件(1)給出的資料表明這是一個等差數列，公差等於-2。第298個數是-616。所以，第293個數 = -616+2x (298-293)，條件(1)滿足。條件(2)只給出了首項，不充分，故答案選A。

11. On a certain date, Hannah invested $5,000 at x percent simple annual interest and a different amount at y percent simple annual interest. What amount did Hannah invest at y percent simple annual interest?

(1) The total amount of interest earned by Hannah’s two investments in one year was $900.

(2) Hannah invested the $5,000 at 6 percent simple annual interest.

題目問：「H在某天以百分之x的年利率投資了$5,000，又以百分之y的年利率投資了另一個數目的錢。問投資Y年利率的錢是多少？」

條件(1)：一年兩種投資賺的錢，總數是$900。

條件(2)：H以6%的年利率，投資了$5,000。

條件(1)給出總收益，但由於x、y均是未知數，所以即使假設另一個數是z，那麼5,000x+zy = 900，還是求不出z是多少。條件(2)已知x等於6，但還是解不出答案。答案選E——兩個條件都不符合。

12. The profit from the sale of a certain appliance increases, though not proportionally, with the number of units sold. Did the profit exceed $4 million on sales of 380,000 units?

(1) The profit exceeded $2 million on sales of 200,000 units.

(2) The profit exceeded $5 million on sales of 350,000 units.

題目問：「某銷售利潤隨著銷售量的增加而增加，但不是按照比例增加。那麼當銷售38萬件貨品時，銷售利潤能超過400萬元嗎？」

條件(1)：當銷售量是20萬件時，利潤超過200萬元。

條件(2)：當銷售量是35萬件時，利潤超過500萬元。

條件(1)給出銷售量是20萬件，利潤200萬元，由於不是按比例增加，所以不能充分表明題幹。條件(2)，由於銷售了35萬件的利潤已經是500萬元了，那麼賣了38萬件，利潤肯定大於500萬元，當然也就超過400萬元，所以條件(2)符合，答案選B。

13. If n is an integer, is n even?

(1) n^2-1 is an odd integer

(2) 3n+4 is an even integer

題目問：「如果n是整數，那麼n是偶數嗎？」

(1) n^2-1是奇數

(2) 3n+4是偶數

條件(1)表明n^2-1是奇數，那麼n^2是偶數，也就是說n是偶數滿足題幹，條件(2)3n+4是偶數，那麼3n是偶數，因為3是奇數，所以n一定是偶數，條件(2)滿足。由於兩個條件都滿足，故答案是D。

14. Mary currently works 30 hours per week at her part-time job. If her gross hourly wage were to increase by $1.50, how many fewer hours could she work per week and still ear n the same gross weekly pay as before the increase?

(1) Her gross weekly pay is currently $225.00.

(2) An increase of $1.50 would represent an increase of 20 percent of her current gross hourly wage.

題目問：「瑪莉在工餘時間打工，打工30個小時，每小時比以前多賺1.5，那麼以前她至少工作多少時間，才能賺得跟現在一樣多？」

條件(1)：目前它的工資是一周$225

條件(2)：長了1.5等於她目前一小時工資的20%

條件(1)已知目前一周賺得總數，一周30小時，所以一小時賺的就知道了，那麼以前一小時賺得也就是到了，這樣就能算出以前需要做多少時間。條件以滿足。條件(2)只能算出目前一小時工資是多少，可以算出之前一小時工資是多少，那麼也就能算出之前需要工作時長。所以連個條件都單獨滿足，答案選D。

15. If 90 students auditioned for the school musical, how many were accepted?

(1) 2/3of the boys and 1/3of the girls who auditioned were accepted.

(2) 26 of the boys who auditioned were accepted.

題目問：「某校有90個學生參加學校試鏡，有多少學生能被選上？」

條件(1)：2/3的男生和1/3的女生被選上

條件(2)：26位男生被選上

由於不知道男、女生的數量，所以條件(1)算不出有多少學生被選上，條件(2)給出男生被選上的數量，那麼就可以知道26佔據男生的2/3，所以男生人數就知道了，這樣女生人數就知道了，然後女生1/3的人數選上，這樣就知道了男女被選上的總人數了所以兩個條件和在一起才能得出答案，故選C。

16. A circular tub has a band painted around its circumference, as shown above. What is the surface area of this painted band?

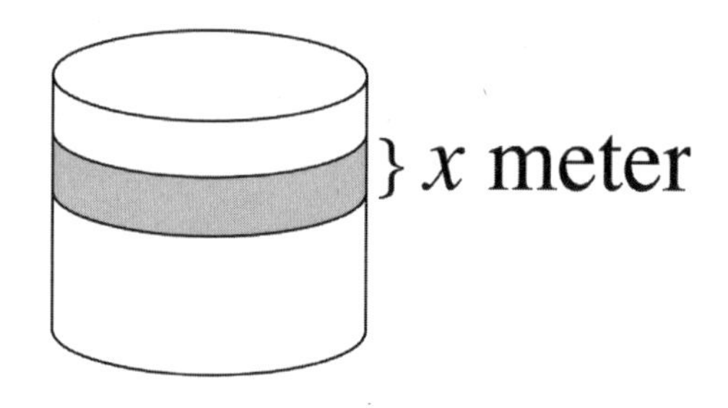

(1) x = 0.5

(2) The height of the tub is 1 meter.

題目問：「一個圓形浴缸，沿著周長塗一圈帶狀面積，求這個帶狀面積是多少？」

條件(1)：x = 0.05

條件(2)：浴缸高度是1米

帶狀面積其實就是一個長方形的面積，只不過長方形高是x，長是圓的周長，條件(1)給出高度　，但是不知道圓的周長，條件(2)的條件也沒有用，故答案選E——兩個條件都不充分。

17. If A and B are positive integers, is the product AB even?

(1) The sum A + B is odd.

(2) A is even.

題目問：「如果A、B是正整數，那麼AB乘積是偶數嗎？」

條件(1)：A、B之和是奇數

條件(2)：A是偶數

AB乘積是偶數，那麼AB中至少一個是偶數就可以。條件(1)得出A、B中有一個是奇數，有一個是偶數，滿足。條件(2)滿足至少一個是偶數。所以兩個條件都能單獨滿足，故答案選D。

18. The weights of all dishes of type X are exactly the same, and the weights of all dishes of type Y are exactly the same. Is the weight of 1 dish of type X less than the weight of 1 dish of type Y?

(1) The total weight of 3 dishes of type X and 2 dishes of type Y is less than the total weight of 2 dishes of type X and 4 dishes of type Y.

(2) The total weight of 4 dishes of type X and 3 dishes of type Y is less than the total weight of 3 dishes of type X and 4 dishes of type Y.

題目問：「X碟的重量均相等，Y碟的重量也均相等，那麼單個X碟的重量比單個Y碟的輕嗎？」

(1) 3個X和2個Y的重量，少於2個X和4個Y。

(2) 4個X和3個Y的重量，少於3個X和4個Y。

條件(1)列出不等式$3X+2Y<2X+4Y$，化簡得出$X<2Y$，條件不夠。

條件(2)列出不等式$4X+3Y<3X+4Y$，化簡得出X。

19. A certain high school with a total enrollment of 900 students held a science fair for three days last week. How many of the students enrolled in the high school attended the science fair on all three days?

(1) Of the students enrolled in the school, 30 percent attended the science fair on two or more days.

(2) Of the students enrolled in the school, 10 percent of those that attended the science fair on at least one day attended on all three days.

題目問：「上周三天某校總共有900位學生參加了科學博覽會，有多少學生三天都參加了？」

條件(1)：有30%的學生參加會議超過兩天

條件(2)：10%至少在了一天

有300人參加會議超過2天，只能說明三天都在的人數少於300，但是不確定是多少人。條件(2)只能說明90個人至少在了一天。還是不能滿足，故答案選E。

20. The inside of a rectangular carton is 48 centimeters long, 32 centimeters wide and 15 centimeters high. The carton is filled to capacity with k identical cylindrical cans of fruit that stand upright in rows and columns, as indicated in the figure above. If the cans are 15 centimeters high. What is the value of k?

(1) Each of the cans has a radius of 4 centimeters.

(2) Six of the cans fit exactly along the length of the carton.

題目問：「長方形紙盒底48厘米，寬32厘米，高15厘米。入邊排列整齊地放滿了相同的K類圓形水果罐頭。如果罐頭的高15厘米，那麼能放多少？」

條件(1)：每個罐頭的半徑是4厘米

條件(2)：6個罐頭可以填滿箱子的一長列

條件(1)每個半徑是4厘米，所以直徑是8厘米，所以相當於長48、寬32的長方形分為多少個長為8的方塊。32/8 = 4，48/8 = 6，所以能放4x6 = 24個。

條件(2)表示6個罐頭可以填滿箱子的一長列，所以算出每個直徑是48/6，所以一寬列能放32/8 = 4個。總共能放24個，所以兩個條件都是單獨滿足，故答案是D。

21. John holds 12 cards, each of which is red, white, green, or blue. If a person is to select a card randomly from the cards Terry is holding, is the probability less than1/2 that the card selected will be either red or white?

(1) The probability that the person will select a blue card is 1/3.

(2) The probability that the person will select a red card is 1/6.

題目問：「約翰手裡有紅、白、綠、藍4種牌共12張，一個人隨即從他手中取出一張牌，那麼取出的牌是紅色或者白色的機會率少於1/2。需要哪些條件？」

條件(1)：藍牌的機率是1/3

條件(2)：紅牌得機率是1/6

條件(1)藍的機率是1/3，那麼紅，白綠有2/3的機率，但是具體紅，白是多少，不確定。條件(2)紅的機率是1/6小於1/2，但是不能確定白色的機率，答案選E。

22. The selling price of an article is equal to the cost of the article plus the markup. The markup on a certain television set is what percent of the selling price?

(1) The markup on the television set is 25 percent of the cost.

(2) The selling price of the television set is $250.

題目問：「一件物品的售價等於物品的成本加上利潤。那麼某電視機的利潤是售價的多少比例？」

條件(1)：利潤是成本的25%

條件(2)：售價是250

條件(1)利潤是成本的25%，假設成本是x，那麼售價就是1.25x，那麼就可以得出0.25/1.25 = 0.2；條件(2)給出售價250，但成本和利潤都不知道是多少，所以不滿足，答案是A。

23. Can a certain rectangular sheet of glass be positioned on a rectangular tabletop so that it covers the entire tabletop and its edges are parallel to the edges of the tabletop?

(1) The tabletop is 36 inches wide by 60 inches long.

(2) The area of one side of the sheet of glass is 2,400 square inches.

題目問：「怎樣才能使一個矩形玻璃板完全覆蓋一個矩形桌子，並且與桌面邊緣平行。」

條件(1)：桌子長60，寬36。

條件(2)：玻璃蓋子一面的面積是2400

只有玻璃蓋子的形狀與長方形桌子相似或全等，才能完全罩住。條件(1)給出桌子長寬，條件(2)給出玻璃蓋子面積，但形狀確定不了。所以兩個條件都不能滿足要求，答案是E。

24. In a random sample of 80 adults, how many are college graduates?

(1) In the sample, the number of adults who are not college graduates is 3 times the number who are college graduates.

(2) In the sample, the number of adults who are not college graduates is 40 more than the number who are college graduates.

題目問：「在一個由80個成年人組成的隨機樣本中，有多少位大學生？」

條件(1)：在隨機樣本中，非大學生是大學生的3倍。

條件(2)：非大學生比大學生的人數多40人

知道大學生和非大學生的關係就能知道具體數量。條件(1)和條件(2)都能單獨滿足，答案是D。

25. The table above shows the distance, in kilometers, by the most direct route, between any two of the four cities, R, S, T, and U. For example, the distance between City R and City U is 62 kilometeRs. What is the value of x?

	R	S	T	U
R	o	y	x	62
S	y	o	56	75
T	x	56	o	69
U	62	75	69	o

(1) By the most direct route, the distance between S and T is twice the distance between R and S.

(2) By the most direct route, the distance between T and U is 1.5 times the distance between R and T.

題目問：「方格裡的數表示兩地之間的直接距離，R和U相距62公里，那麼X是多少？」

條件(1)：S和T之間的直線距離，是R和S之間直線距離的2倍。

條件(2)：T和U之間的直線距離，是R和T之間直線距離的1.5倍。

知道四個地點的位置就能得出X的數值。X表示R、T之間的距離，S、T之間的直線距離是S、R之間直線距離的2倍，那麼56 = 2y，得出y，但X不知道條件(2)1.5x = 69可以求出X，所以答案是B。

26. What is the tenths digit in the decimal representation of a certain number?

(1) The number is less than 1/3

(2) The number is greater than 1/4

題目問：「某個十進制數的十分位數字是多少？」

條件(1)：這個數少於1/3

條件(2)：這個數大於1/4

少於1/3，十分位可以是0、1、2、3，大於1/4，可以是2、3……9。還是不能確定唯一，答案是E。

試題練習

1. A dice is rolled randomly on to a circular board with a triangle inscribed in the circle (All three vertices of the triangle are on the circumference of the circle).What is the probability that the die comes to rest outside the triangular region?

(1) The hypotenuse of the triangle is a diameter of the circle.

(2) The radius of the circle is 2 units, and the area of the triangle is 4 square units.

A. Statement (1) alone is sufficient, but statement (2) alone is not sufficient to answer the question.

B. Statement (2) alone is sufficient, but statement (1) alone is not sufficient to answer the question.

C. Both statements taken together are sufficient to answer the question, but neither statement alone is sufficient.

D. Each statement alone is sufficient.

E. Statements (1) and (2) together are not sufficient, and additional data is needed to answer the question.

2. In what year was Mary born?

(1) Mary's daughter was born in 1960 when Mary was 28 years old.

(2) Mary's birthday and her daughter's birthday are exactly six months apart.

A. Statement (1) alone is sufficient, but statement (2) alone is not sufficient to answer the question.

B. Statement (2) alone is sufficient, but statement (1) alone is not sufficient to answer the question.

C. Both statements taken together are sufficient to answer the question, but neither statement alone is sufficient.

D. Each statement alone is sufficient.

E. Statements (1) and (2) together are not sufficient, and additional data is needed to answer the question.

3. Is $xy < 15$?

(1) $0.5 < x < 1$, and $y^2 = 144$

(2) $x < 3$, $y < 5$

A. Statement (1) alone is sufficient, but statement (2) alone is not sufficient to answer the question.

B. Statement (2) alone is sufficient, but statement (1) alone is not sufficient to answer the question.

C. Both statements taken together are sufficient to answer the question, but neither statement alone is sufficient.

D. Each statement alone is sufficient.

E. Statements (1) and (2) together are not sufficient, and additional data is needed to answer the question.

4. $x = 1.24d5$

If d is the thousandth's digit in the decimal above, what is the value of x when rounded to the nearest hundredth?

(1) $x < \frac{5}{4}$

(2) $d < 5$

A. Statement (1) alone is sufficient, but statement (2) alone is not sufficient to answer the question.

B. Statement (2) alone is sufficient, but statement (1) alone is not sufficient to answer the question.

C. Both statements taken together are sufficient to answer the question, but neither statement alone is sufficient.

D. Each statement alone is sufficient.

E. Statements (1) and (2) together are not sufficient, and additional data is needed to answer the question.

5. Are the integers x, y and z consecutive?

(1) The arithmetic mean (average) of x, y and z is y.

(2) $y-x = z-y$

A. Statement (1) alone is sufficient, but statement (2) alone is not sufficient to answer the question.

B. Statement (2) alone is sufficient, but statement (1) alone is not sufficient to answer the question.

C. Both statements taken together are sufficient to answer the question, but neither statement alone is sufficient.

D. Each statement alone is sufficient.

E. Statements (1) and (2) together are not sufficient, and additional data is needed to answer the question.

6. How many ewes (female sheep) in a flock of 50 sheep are black?

(1) There are 10 rams (male sheep) in the flock.

(2) 40% of the animals are black.

A. Statement (1) alone is sufficient, but statement (2) alone is not sufficient to answer the question.

B. Statement (2) alone is sufficient, but statement (1) alone is not sufficient to answer the question.

C. Both statements taken together are sufficient to answer the question, but neither statement alone is sufficient.

D. Each statement alone is sufficient.

E. Statements (1) and (2) together are not sufficient, and additional data is needed to answer the question.

7. If a and b are both positive, what percent of b is a?

(1) a = 3/11

(2) b/a = 20

A. Statement (1) alone is sufficient, but statement (2) alone is not sufficient to answer the question.

B. Statement (2) alone is sufficient, but statement (1) alone is not sufficient to answer the question.

C. Both statements taken together are sufficient to answer the question, but neither statement alone is sufficient.

D. Each statement alone is sufficient.

E. Statements (1) and (2) together are not sufficient, and additional data is needed to answer the question.

8. A wheel of radius 2 meters is turning at a constant speed. How many revolutions does it make in time T?

(1) T = 20 minutes

(2) The speed at which a point on the circumference of the wheel is moving is 3 meters per minute.

A. Statement (1) alone is sufficient, but statement (2) alone is not sufficient to answer the question.

B. Statement (2) alone is sufficient, but statement (1) alone is not sufficient to answer the question.

C. Both statements taken together are sufficient to answer the question, but neither statement alone is sufficient.

D. Each statement alone is sufficient.

E. Statements (1) and (2) together are not sufficient, and additional data is needed to answer the question.

9. Is $x > 0$?

(1) $-2x < 0$

(2) $x^3 > 0$

A. Statement (1) alone is sufficient, but statement (2) alone is not sufficient to answer the question.

B. Statement (2) alone is sufficient, but statement (1) alone is not sufficient to answer the question.

C. Both statements taken together are sufficient to answer the question, but neither statement alone is sufficient.

D. Each statement alone is sufficient.

E. Statements (1) and (2) together are not sufficient, and additional data is needed to answer the question.

10. A certain straight corridor has four doors, A, B, C and D (in that order) leading off from the same side. How far apart are doors B and C?

(1) The distance between doors B and D is 10 meters.

(2) The distance between A and C is 12 meters.

A. Statement (1) alone is sufficient, but statement (2) alone is not sufficient to answer the question.

B. Statement (2) alone is sufficient, but statement (1) alone is not sufficient to answer the question.

C. Both statements taken together are sufficient to answer the question, but neither statement alone is sufficient.

D. Each statement alone is sufficient.

E. Statements (1) and (2) together are not sufficient, and additional data is needed to answer the question.

11. Two socks are to be picked at random from a drawer containing only black and white socks. What is the probability that both are white?

(1) The probability of the first sock being black is 1/3.

(2) There are 24 white socks in the drawer.

A. Statement (1) alone is sufficient, but statement (2) alone is not sufficient to answer the question.

B. Statement (2) alone is sufficient, but statement (1) alone is not sufficient to answer the question.

C. Both statements taken together are sufficient to answer the question, but neither statement alone is sufficient.

D. Each statement alone is sufficient.

E. Statements (1) and (2) together are not sufficient, and additional data is needed to answer the question.

12. A bucket was placed under a dripping tap which was dripping at a uniform rate.At what time was the bucket full?

(1) The bucket was put in place at 2pm.

(2) The bucket was half full at 6pm and three-quarters full at 8pm on the same day.

A. Statement (1) alone is sufficient, but statement (2) alone is not sufficient to answer the question.

B. Statement (2) alone is sufficient, but statement (1) alone is not sufficient to answer the question.

C. Both statements taken together are sufficient to answer the question, but neither statement alone is sufficient.

D. Each statement alone is sufficient.

E. Statements (1) and (2) together are not sufficient, and additional data is needed to answer the question.

13. What is the ratio of male to female officers in the police force in town T?

(1) The number of female officers is 250 less than half the number of male officers.

(2) The number of female officers is 1/7 the number of male officers.

A. Statement (1) alone is sufficient, but statement (2) alone is not sufficient to answer the question.

B. Statement (2) alone is sufficient, but statement (1) alone is not sufficient to answer the question.

C. Both statements taken together are sufficient to answer the question, but neither statement alone is sufficient.

D. Each statement alone is sufficient.

E. Statements (1) and (2) together are not sufficient, and additional data is needed to answer the question.

14. How long did it take Henry to drive to work last Wednesday? (He did not stop on the way).

(1) If he had driven twice as fast he would have taken 35 minutes.

(2) His average speed was 30 miles per hour.

A. Statement (1) alone is sufficient, but statement (2) alone is not sufficient to answer the question.

B. Statement (2) alone is sufficient, but statement (1) alone is not sufficient to answer the question.

C. Both statements taken together are sufficient to answer the question, but neither statement alone is sufficient.

D. Each statement alone is sufficient.

E. Statements (1) and (2) together are not sufficient, and additional data is needed to answer the question.

15. What is the slope of line l which passes through the origin of a rectangular coordinate system?

(1) The line does not intersect with the line $y = x + 2$

(2) The line passes through the point (3, 3)

A. Statement (1) alone is sufficient, but statement (2) alone is not sufficient to answer the question.

B. Statement (2) alone is sufficient, but statement (1) alone is not sufficient to answer the question.

C. Both statements taken together are sufficient to answer the question, but neither statement alone is sufficient.

D. Each statement alone is sufficient.

E. Statements (1) and (2) together are not sufficient, and additional data is needed to answer the question.

16. Fifty percent of the articles in a certain magazine are written by staff membeRs.Sixty percent of the articles are on current affaiRs.If 75 percent of the articles on current affairs are written by staff members with more than 5 years experience of journalism, how many of the articles on current affairs are written by staff members with more than 5 years experience?

(1) 20 articles are written by staff members.

(2)Of the articles on topics other than current affairs, 50 percent are by staff members with less than 5 years' experience.

A. Statement (1) alone is sufficient, but statement (2) alone is not sufficient to answer the question.

B. Statement (2) alone is sufficient, but statement (1) alone is not sufficient to answer the question.

C. Both statements taken together are sufficient to answer the question, but neither statement alone is sufficient.

D. Each statement alone is sufficient.

E. Statements (1) and (2) together are not sufficient, and additional data is needed to answer the question.

17. Is $xy > 0$?

(1) $\frac{x}{y} < 0$

(2) $x + y < 0$

A. Statement (1) alone is sufficient, but statement (2) alone is not sufficient to answer the question.

B. Statement (2) alone is sufficient, but statement (1) alone is not sufficient to answer the question.

C. Both statements taken together are sufficient to answer the question, but neither statement alone is sufficient.

D. Each statement alone is sufficient.

E. Statements (1) and (2) together are not sufficient, and additional data is needed to answer the question.

18. One number, n, is selected at random from a set of 10 integeRs.What is the probability that $\frac{1}{2}n + 13 = 0$?

(1) The largest integer in the set is 13

(2) The arithmetic mean of the set is zero.

A. Statement (1) alone is sufficient, but statement (2) alone is not sufficient to answer the question.

B. Statement (2) alone is sufficient, but statement (1) alone is not sufficient to answer the question.

C. Both statements taken together are sufficient to answer the question, but neither statement alone is sufficient.

D. Each statement alone is sufficient.

E. Statements (1) and (2) together are not sufficient, and additional data is needed to answer the question.

19. Is w an integer?

(1) 3w is an odd number

(2) 2w is an even number

A. Statement (1) alone is sufficient, but statement (2) alone is not sufficient to answer the question.

B. Statement (2) alone is sufficient, but statement (1) alone is not sufficient to answer the question.

C. Both statements taken together are sufficient to answer the question, but neither statement alone is sufficient.

D. Each statement alone is sufficient.

E. Statements (1) and (2) together are not sufficient, and additional data is needed to answer the question.

20. A piece of string $1\frac{1}{2}$ meters long is cut into three pieces. What is the length of each of the pieces?

(1) The length of one of the pieces is 20 cm.

(2) The sum of the lengths of two of the pieces is equal to the length of the third piece.

A. Statement (1) alone is sufficient, but statement (2) alone is not sufficient to answer the question.

B. Statement (2) alone is sufficient, but statement (1) alone is not sufficient to answer the question.

C. Both statements taken together are sufficient to answer the question, but neither statement alone is sufficient.

D. Each statement alone is sufficient.

E. Statements (1) and (2) together are not sufficient, and additional data is needed to answer the question.

答案及解析

1. Answer: B

The find the probability we need to know the ratio of the area inside the triangle to the area outside. Statement (1) tells us that the triangle lies in one semi-circle and is a right angled triangle, but does not allow us to fix the exact area of the triangle. Thus the answer cannot be A or D. Statement (2) allows us to find the area of the circle and the area of the triangle is given, and so is sufficient to find the ratio of the areas and the answer is B.

2. Answer: E

Someone who is 28 in one particular year could have become 28 that same year or have become 28 the previous year, depending on the month of birth, and so the information in statement (1) is not sufficient to answer the question of whether Mary was born in 1931 or 1932. The information in statement (2) places the birthdays six months apart, but give no information on age hence the answer must be C or E. Even combined with statement (1) the information in statement (2) leaves open the possibility of a 1931 or 1932 birth year for Mary and so the answer is E.

3. Answer: A

From statement (1) we can see that y must be 7 or -7 and x is a positive fraction. A fraction of 7 or of -7 will always be less than 15 and so statement (1) is sufficient and the answer must be A or D. Statement (2) alone does not answer the questions definitely because x and y could both be positive numbers with a product less than 15 of they could both be negative with large absolute values so that their product is greater than 15. The answer is A.

4. Answer: D

When rounded to the nearest hundredth x can be 1.24 or 1.25, depending on whether we round up or down. If d is ≥ 5 the value of x when rounded up will be 1.25. Statement (1) says that x is less than 1.25. Hence we know we must round down to 1.24 and the answer must be A or D. If d is less than 5, as stated in statement (2), we know we have to round down to 1.24 and so the answer is D.

5. Answer: E

The mean of three numbers will equal the middle number for any set of evenly spaced numbers (1, 2, 3 or 2, 4, 6, or -1, -4, -7 for example) and so we cannot assume that x, y and z are consecutive. Hence the answer cannot be A or D. If x, y and z were 2, 4, and 6, for example, the equation in statement (2) would be valid, so once again the numbers do not have to be consecutive, and the answer cannot be B. From the numbers we have just substituted, we should be able to see that putting statements (1) and (2) together will still not give a situation in which the numbers are always consecutive or always not consecutive, and so the best answer is E.

6. Answer: E
From statement (1) we know the ratio of male to female sheep, but nothing about the color distribution. So the answer cannot be A or D.
From statement (2) we know that 40% of the animals are black but nothing about whether they are male of female. So the answer cannot be B. Even putting the information together does not help because there is no way to tell what fraction of the female sheep are black. And so C cannot be correct, and the answer is E.

7. Answer: B
Statement (1) tells us nothing about b and so the answer cannot be A or D. To find what percent a is of b we need to solve the expression a/b x 100.
Statement (2) allows us to do just that: a/b = 1/20. No need to go any further; the answer is B.

8. Answer: C
To find the number of revolutions we need to know the rate of turning and the time duration.
Statement (1) gives us only the time, and so the answer cannot be A or D.
Statement (2) tells us the rate at which a point on the circumference is moving, which, since we know the dimensions of the wheel, is sufficient to determine the number of rotations per minute. But since we do not know the time, B cannot be correct. But putting statements (1) and (2) together we have all we need, so the answer is C.

9. Answer: D
The statement that x is greater than zero means that x is positive. If we multiply a posi-

tive number by a negative number the product is negative: this is what we get from statement (1), which thus tells us that x is positive. The answer must be A or D. The cube of a positive number is positive; the cube of a negative number is negative, and so statement (2) tells us that x is positive. And so the answer is D.

10. Answer: E

It is obvious that neither statements (1) or (2) alone can tell you how far apart B and C are, and so the answer must be C or E. To see whether putting both pieces of information together will be adequate, visualize two rods: BD of length 10 units, and AC of length 12 units. Mentally place the rods alongside each other so that C lies between B and D. Now you can mentally slide the rods past each other to see that C can lie anywhere between B and D, and so we cannot fix one value for the length BC, and the answer is E.

11. Answer: C

From statement (1) we know the ratio of black socks to white, but that ratio will change when one sock is taken out. To get the new ratio, and hence the probability that the next sock will also be white, we need to know the number of socks of each type. The answer cannot be A or D. Obviously statement (2) on its own does not get the ratio and so B cannot be correct. But putting the information in both statements together we can solve the problem (24 white socks with a ratio of black to total of 1:3 means that there are 12 black and 24 white socks). The answer is C.

12. Answer: B

Since we need rate of dripping, statement (1) is not enough and the answer cannot be A or D. Ignoring (1) and looking at statement (2) we can easily solve the problem because one quarter of the bucket got filled in 2 hours and the filling will get over at 10pm. The answer is B.

13. Answer: B

To get the ratio we need total numbers male to total numbers female. There is not enough information in statement (1), so A and D cannot be correct. A ratio is just a fraction, so if we have the fraction of female officers we have the ratio (assuming that all officers have to be either male or female!!!). So the answer is B.

14. Answer: A

Do not overcomplicate this one! From statement (1) we can directly say that if he had driven at his normal speed he would have taken twice the time (i.e. 70 minutes). So the answer must be A or D. Looking at statement (2) alone we have speed but not distance and so we cannot get time. The answer is A.

15. Answer: D

If a straight line passes through the origin we only need one other point to fix the line. And so we should note that statement (2) gives us one point and the answer must be B or D. Considering statement (1) alone, if two lines do not intersect, they must be parallel (have the same slopes). In the equation for a straight line $y = mx + c$, m is the slope. In this case $m = 1$ and we have answered the question. The answer is D.

16. Answer: A

There is only one item of information in statement (1), but we can combine it with the given information to find that the total number of articles is 40. Further we can see that 24 are on current affairs, of which 18 are by experienced staff membeRs.So the answer must be A or D. Ignoring statement (1), we can see that the information in statement (2) gives us a fraction and not a number and no amount of trying will get us one definite number from which we can find a discrete answer. Hence the answer is A.

17. Answer: A

If xy is greater than zero, then either both x and y must be negative, or both must be positive. In statement (1) we have x/y shown as a negative fraction, so either x or y (but not both) must be negative. Hence xy cannot be positive. Since we have a definite solution, the answer must be A or D. Looking at statement (2) alone, we can see that x and y could both be negative, but it is also possible that one could be negative and the other positive depending on the absolute values of the numbeRs.Hence the answer is A.

18. Answer: E

Statement (1) alone tells us the largest number in the set. If that happened to be the number picked, then we would be certain that $1n/2 + 13$ was not less than or equal to zero. However, we have insufficient information on the other numbers in the set. (If for example -26 was in the set, and the number picked happened to be -26, then we are certain that the value of the expression would be zero.) And so the answer cannot be A or D.

From statement (2) alone we know that some of the numbers must be positive and some negative, but once again, we have no information on the actual numbeRs.(Do not assume that the smallest number has to be -13). Thus B cannot be the answer. Combining the information is also not sufficient to get the answer because we do not know what the smallest number is. The answer is E.

19. Answer: B

If w happened to be 1, then 3w would be an odd number, but we can pick a fraction such as 5/3 for w which also makes 3w an odd number. So the answer cannot be A or D. However, in statement (2) we are told that twice the number is even. All even numbers when divided by 2 will give us whole numbers, and so the answer is B.

20. Answer: C

Considering statement (1) we cannot get information on the other two parts of the string. Thus the answer cannot be A or D. From statement (2) we can write the equation $x + y = z$, and we know that $x + y + z = 150$, but we cannot get any further, and so the answer cannot be B. Now if we take the information in both statements we can see that 20cm cannot be the length of the longest piece (z) and so in our equations we can call it x or y. Now our equations can be solved because we have two equations and two unknowns. ($x + 20 = z$; and $x + 20 + z = 150$. hence $x + 20 + x + 20 = 150$, and $x = 55$). There was no need to solve. Also we should have spotted that since x + y is equal to z, then z is half the string (75cm) and x + y is the other half, so once we have either x or y, we have the solution. Hence answer C.

PART FOUR

Numerical Reasoning

試前準備

1. 熟記各種數字的運算關係

如各個數字的平方、立方以及與它們的相鄰數字，以便做到當一看到某個數字時就有「感覺」。這是迅速準確解好數字推理題材的前提。常見而需記住的數字關係如下：

(1)平方關係：1(1^2)、4(2^2)、9(3^2)、16(4^2)、25(5^2)，如此類推。

(2)立方關係：8(2^3)、27(3^3)、64(4^3)、125(5^3)，如此類推。

(3)質數關係：2、3、5、7、11、13等

(4)開方關係：諸如2($\sqrt{4}$)、3($\sqrt{9}$)、4($\sqrt{16}$)、5($\sqrt{25}$)

以上幾種（特別是前兩種關係），每次考試必有。所以，對這些平方立方後的數字，及這些數字的鄰居（如64、63、65等）要有足夠的敏感。當看到這些數字時，立刻就能想到平方或立方的可能性。

2.　熟練掌握各種簡單運算，一般加減乘除大家都會，值得注意的是帶根號的運算。根號運算掌握簡單規律則可，也不難。

3.　對於中等難度以下的題目，建議大家練習使用心算，可節省不少時間，在考試時有很大效果。

答題技巧

按數字之間的關係，可將數字推理題分為下列8類：

1. 和差關係

和差關係可細分為「等差關係」和「移動求和或差」兩種：

(1)等差關係：這種題屬於比較簡單的，不經練習也能在短時間內做出。

例題：

【1】12，20，30，42，( )

答案：56。由於12+8 = 20，20+10 = 30，30+12 = 42，故此最後的未知數的數值，應跟42相差14。

【2】127，112，97，82，( )

答案：67。數值間的差距為15。

【3】3，4，7，12，( )，28

答案：19。數值間的差距(順序)為1，3，5，7，9。28-9 = 19。

(2)移動求和或差：從第三項起，每一項都是前兩項之和或差。雖然這種題初次做稍有難度，做多了也就簡單了。

例題：

【1】1，2，3，5，( )，13

答案：8。1+2 = 3，2+3 = 5，3+5 = 8，5+8 = 13。

【2】2，5，7，( )，19，31，50

答案：12。2+5 = 7，5+7 = 12，7+12 = 19，19+31 = 50。

【3】5，3，2，1，1，( )

答案：0。5-3 = 2，3-2 = 1，2-1 = 1，1-1 = 0。

2. 乘除關係

乘除關係又分為「等比關係」和「移動求積或商關係」兩種：

(1)等比關係：從第二項起，每一項與它前一項的比等於一個常數或一個等差數列。

例題：

【1】8，12，18，27，( )

答案：40.5。後項與前項之比為1.5倍。

【2】6，6，9，18，45，( )

答案：135。後項與前項之比為等差數列，分別為1、1.5、2、2.5、3倍。

(2)移動求積或商關係。從第三項起，每一項都是前兩項之積或商。

例題：

【1】2，5，10，50，( )

答案：500。2x5 = 10，5x10 = 50，10x50 = 500。

【2】3，4，6，12，36，( )

答案：216。此題稍有難度，從第三項起，第項為前兩項之積除以2。

【3】1，7，8，57，( )

答案：457。後項為前兩項之積+1。

3. 平方關係

「平方」是指兩個相同數的乘積是這個數的平方。如4就是2的平方。

例題：

【1】1，4，9，16，25，( )，49

答案：36。括號前的幾個數字(1、4、9、16和25)其實是1、2、3、4和5的平方，而49則是7的平方，故估計括號內的數字是6的平方，即36。

【2】66，83，102，123，( )

答案：146。因數列的數值為8，9，10，11，12的平方後+2。

4. 立方關係

「立方」是指3個相同數的乘積是這個數的立方。如8是2的立方。

例題：

【1】1，8，27，( )，125

答案：81。上列的數值為1、2、3、4和5的立方。

【2】3，10，29，( )，127

答案：83。上列的數值為立方後+2。

5. 分數數列

一般這種數列出難題較少，關鍵是把分子和分母看作兩個不同的數列，有的還需進行簡單的通分，則可得出答案。

例題：

【1】1/2，4/3，9/4，16/5，25/6，( )

答案：36/7。分子為1至5的平方數，至於分母是以「+1」的形式遞增。

6. 帶根號的數列

這種題難度一般也不大，掌握根號的簡單運算則可。限於計算機水平比較爛，打不出根號，無法列題。

7. 雙重數列

雙重數列關係可分為三種，例如：

(1)每兩項為一組，如：

例題：

【1】1，3，3，9，5，15，7，( )

答案：21。第一與第二、第三與第四等每兩項後項與前項之比為3。

【2】2，5，7，10，9，12，10，( )

答案：13。每兩項之差為3

【3】1/7，14，1/21，42，1/36，72，1/52，( )

答案：104。兩項為一組，每組的後項等於前項倒數乘2。

(2)兩個數列相隔，其中一個數列可能無任何規律，但只要把握有規律變化的數列就可得出結果。

例題：

【1】22，39，25，38，31，37，40，36，( )

答案：52。由兩組數列「22，25，31，40，(　　)」和「39，38，37，36」組成，相互隔開，均為等差。

【2】34，36，35，35，(36)，34，37，( )

答案：33。由兩個數列相隔而成，一個遞增，一個遞減。

(3)數列中的數字帶小數，其中整數部分為一個數列，小數部分為另一個數列。

例題：

【1】2.01，4.03，8.04，16.07，( )

答案：32.11。整數部分為等比，小數部分為移動求和數列。雙重數列難題也較少。能看出是雙重數列，題目一般已經解出。特別是前兩種，當數字的個數超過7個時，為雙重數列的可能性相當大。

8. 組合數列

此種數列最難。前面8種數列，單獨出題幾乎沒有難題，也出不了難題，但8種數列關係兩兩組合，變態的甚至三種關係組合，就形成了比較難解的題目了。最常見的是和差關係與乘除關係組合、和差關係與平方立方關係組合。只有在熟悉前面所述8種關係的基礎上，才能較好較快地解決這類題。

例題：

【1】1，1，3，7，17，41，( )

答案：99。此為移動求和與乘除關係組合。第三項 = (第二項)x2+第一項。

【2】65，35，17，3，( )

答案：1。平方關係與和差關係組合，分別為8^2+1，6^2-1，4^2+1，2^2-1，下一個應為$0^2+1 = 1$。

【3】4，6，10，18，34，( )

答案：66。各差關係與等比關係組合。依次相減，得2，4，8，16，( )，可推知下一個為32，32+34 = 66。

【4】6，15，35，77，( )

答案：163。等差與等比組合。前項x2+3，5，7依次得後項，得出下一個應為77x2+9 = 163。

【5】0，6，24，60，120，( )

答案：210。和差與立方關係組合。$0 = 1^3-1$，$6 = 2^3-2$，$24 = 3^3-3$，$60 = 4^3-4$，$120 = 5^3-5$。

【6】1，4，8，14，24，42，( )

答案：76。兩個等差與一個等比數列組合，依次相減得出3，4，6，10，18，( )。再相減，得出1，2，4，8，( )，此為等比數列，下一個為16，倒推可知選A。

這些數列部分也屬於組合數列，但由於與前面所講的和差、乘除、平方等關係不同，故在此列為其他數列。這種數列一般難題也較多。

試題練習

【1】7，9，-1，5，( )

A. 4

B. 2

C. -1

D. -3

【2】3，2，5/3，3/2，( )

A. 1/4

B. 7/5

C. 3/4

D. 2/5

【3】1，2，5，29，( )

A. 34

B. 841

C. 866

D. 37

【4】2，12，30，( )

A. 50

B. 65

C. 75

D. 56

【5】2，1，2/3，1/2，( )

A. 3/4

B. 1/4

C. 2/5

D. 5/6

【6】4，2，2，3，6，( )

A. 6

B. 8

C. 10

D. 15

【7】 1，7，8，57，( )

A. 123

B. 122

C. 121

D. 120

【8】 4，12，8，10，( )

A. 6

B. 8

C. 9

D. 24

【9】 1/2，1，1，( )，9/11，11/13

A. 2

B. 3

C. 1

D. 7/9

【10】95，88，71，61，50，( )

A. 40

B. 39

C. 38

D. 37

【11】2，6，13，39，15，45，23，( )

A. 46

B. 66

C. 68

D. 69

【12】5，15，10，215，( )

A. 415

B. -115

C. 445

D. -112

【13】0，1，3，10，( )

A. 101

B. 102

C. 103

D. 104

【14】124，3612，51020，( )

A. 7084

B. 71428

C. 81632

D. 91836

【15】1，1，2，6，24，( )

A. 25

B. 27

C. 120

D. 125

【16】3，4，8，24，88，( )

A. 121

B. 196

C. 225

D. 344

【17】20，22，25，30，37，( )

A. 48

B. 49

C. 55

D. 81

【18】1/9，2/27，1/27，( )

A. 4/27

B. 7/9

C. 5/18

D. 4/243

【19】1，3，4，8，16，( )

A. 26

B. 24

C. 32

D. 16

【20】2，1，2/3，1/2，( )

A. 3/4

B. 1/4

C. 2/5

D. 5/6

【21】1，1，3，7，17，41，( )

A. 89

B. 99

C. 109

D. 119

【22】6，15，35，77，( )

A. 106

B. 117

C. 136

D. 163

【23】1，3，3，6，7，12，15，( )

A. 17

B. 27

C. 30

D. 24

【24】2/3，1/2，3/7，7/18，( )

A. 4/11

B. 5/12

C. 7/15

D. 3/16

【25】63，26，7，0，-2，-9，( )

A. -16

B. -25

C. -28

D. -36

【26】1，2，3，6，11，20，( )

A. 25

B. 36

C. 42

D. 37

【27】1，2，3，7，16，( )

A. 66

B. 65

C. 64

D. 63

【28】2，6，12，20，( )

A. 40

B. 32

C. 30

D. 28

【29】0，6，24，60，120，( )

A. 186

B. 210

C. 220

D. 226

【30】1，2，3，6，12，( )

A. 16

B. 20

C. 24

D. 36

【31】-2，-8，0，64，( )

A. -64

B. 128

C. 156

D. 250

【32】0，4，18，48，100，( )

A. 140

B. 160

C. 180

D. 200

【33】65，35，17，3，( )

A. 1

B. 2

C. 0

D. 4

【34】1，6，13，( )，33

A. 22

B. 21

C. 20

D. 19

【35】48，2，4，6，54，( )，3，9

A. 6

B. 5

C. 2

D. 3

【36】120，20，( )，-4

A. 0

B. 16

C. 18

D. 19

【37】 -7，3，4，( )，11

A. -6

B. 7

C. 10

D. 13

【38】 3.3，5.7，13.5，( )

A. 7.7

B. 4.2

C. 11.4

D. 6.8

【39】 33.1，88.1，47.1，( )

A. 29.3

B. 34.5

C. 16.1

D. 28.9

【40】5，12，24，36，48，( )

A. 58

B. 60

C. 62

D. 72

【41】16，25，36，50，81，100，169，200，( )

A. 289

B. 225

C. 324

D. 441

答案及解析

1. 答案：D。將數字相加得出：
7+9 = 16
9+(-1) = 8
(-1)+5 = 4
5+(-3) = 2
相加數16、8、4、2屬等比關係。

2. 答案：B。數列可化為：3/1，4/2，5/3，6/4。
分子：3、4、5、6
分母：1、2、3、4
由於兩者均呈遞增關係，故答案是7/5。

3. 答案：C。5 = 1^2+2^2；29 = 5^2+2^2；故( ) = 29^2+5^2，即866。

4. 答案：D。1×2 = 2；3×4 = 12；5×6 = 30；7×8 = ( ) = 56。

5. 答案：C。數列可化為：4/2，4/4，4/6，4/8，分子都是4，分母2、4、6、8則是等差關係，故後項為4/10，即2/5。

6. 答案：D。數列可化為：2/4 = 0.5；2/2 = 1；3/2 = 1.5；6/3 = 2；
由於0.5，1，1.5，2屬等比關係，故後項為2.5×6 = 15。

7. 答案：C。1^2+7 = 8；7^2+8 = 57；8^2+57 = 121。

8. 答案：C。(4+12) / 2 = 8；(12+8) / 2 = 10；(8+10) / 2 = 9。

9. 答案：C。將算式化成：1/2，3/3，5/5，( )，9/11，11/13，這下就看出來了只能是7/7。注意分母是質數列，分子是奇數列。

10. 答案：A。
思路一：它們的十位是一個遞減數字9、8、7、6、5。只是少開始的4，所以選擇A。
思路二：95-9-5 = 81，88-8-8 = 72，71-7-1 = 63，61-6-1 = 54，50-5-0 = 45，40-4-0 = 36，構成等差數列。

11. 答案：D。將數字分為兩個為一組，你會發現後一個數字，其實是前一個數字的3倍。

12. 答案：B。將(前一項)2-(後一項) = 第三項，
例如：5^2-15 = 10，15^2-10 = 215，10^2-215 = -115。

13. 答案：B。0×0+1 = 1，1×1+2 = 3，3×3+1 = 10，10×10+2 = 102。

14. 答案：B。
124可拆成1、2、4
3612可拆成3、6、12
51020可拆成5、10、20
71428可拆成7、14、28
每列都成等差。

15. 答案：C。(1+1)×1 = 2，(1+2)×2 = 6，(2+6)×3 = 24，(6+24)×4 = 120。

16. 答案：D。4 = 2^0+3，8 = 2^2 +4，24 = 2^4 +8，88 = 2^6+24，344 = 2^8+88。

17. 答案：A。兩項相減：2、3、5、7、11，屬質數列的關係。

18. 答案：D。將1/9，2/27，1/27轉化成 1/9，2/27，3/81。
分子1、2、3，屬等差關係；
分母9、27、81，屬等比關係；
故答案是4/243。

19.答案：C。每項都等於其前所有項的和1+3 = 4，1+3+4 = 8，1+3+4+8 = 16，1+3+4+8+16 = 32。

20. 答案：C。將2，1， 2/3，1/2化成2/1，2/2，2/3，2/4。
分子都為2，
分母為1、2、3、4的等差順序，
故括號內的數值為2/5。

21. 答案：B。從第三項開始，都等於：(前一項)x2+再前一項，即：
3 = 1x2+1 = 3
7 = 3x2+1
17 = 7x2+3
41 = 17x2+7
故此，2×41+17 = 99。

22. 答案：D。15 = 6×2+3；35 = 15×2+5；77 = 35×2+7；163 = 77×2+9其中3、5、7、9為等差。

23. 答案：D。1，3，3，6，7，12，15，(24)。奇數項1、3、7、15，新的數列相鄰兩數的差為2、4、8 作差，屬等比關係，至於偶數項3、6、12、24亦同屬等比關係。

24. 答案：A。將數列轉化成：4/6，5/10，6/14，7/18。
分子是：4、5、6、7，故接下來應該是8。
分母是：6、10、14、18，是「加4」的關係，故接下來是22。
再將8/22化簡，得出4/11。

25. 答案：C。4^3-1 = 63；3^3-1 = 26；2^3-1 = 7；1^3-1 = 0；$(-1)^3$-1 = -2；$(-2)^3$-1 = -9；$(-3)^3$-1 = -28。

26. 答案：D。(第一項+第二項+第三項) = 第四項：6+11+20 = 37。

27. 答案：B。前項的平方加後項等於第三項。

28. 答案：C。2 = 2^2-2；6 = 3^2-3；12 = 4^2-4；20 = 5^2-5；30 = 6^2-6。

29. 答案：B。0 = 1^3-1，6 = 2^3-2，24 = 3^3-3，60 = 4^3-4，120 = 5^3-5，210 = 6^3-6

30. 答案：C。按數列上數字的先後次序，分成三組：
第一組：1，2
第二組：3，6
第三組：12，(24)
每組後項除以前項2、2、2。

31. 答案：D。$1^3\times(-2)=-2$；$2^3\times(-1)=-8$；$3^3\times0=0$；$4^3\times1=64$；所以$5^3\times2=250$，故選D。

32. 答案：C。$0=1^0\text{x}0$，$4=2^2\times1$，$18=3^2\times2$，$48=4^2\times3$，$100=5^2\times4$，$180=6^2\times5$。

33. 答案：A。$65=8\times8+1$；$35=6\times6-1$；$17=4\times4+1$；$3=2\times2-1$；$1=0\times0+1$。

34. 答案：A。後一項的數跟前一項的數以5、7、9、11的關係遞增。

35. 答案：C。按數列上數字的先後次序，分成兩組：
第一組：48，2，4，6
第二組：54，( )，3，9。
其中，每組後三個數相乘等於第一個數：$48=2\times4\times6$，而$54=2\times3\times9$。

36. 答案：A。$120=5^3-5$；$20=5^2-5$；$0=5^1-5$；$-4=5^0-5$。

37. 答案：B。前兩個數相加的和的絕對值，等於第三個數：$-7+3=4$；$3+4=7$；$4+7=11$，故選B。

38. 答案：A。小數點的左邊3、5、13、7都為奇數，小數點右邊3、7、5、7同為奇數。考生當遇到數列中所有數都是小數的題時，先不要考慮運算關係，而是直接觀察數字本身，往往數字本身是切入點。

39. 答案：C。
小數點的左邊：33、88、47、16屬「奇、偶、奇、偶」的規律；
小數點右邊：1、1、1、1的等差關係。

40. 答案：B。$12=2\times5+2$；$24=4\times5+4$；$36=6\times5+6$；$48=8\times5+8$。故( )內的數值為$10\text{x}5+10$。

41. 答案：C。奇數項：16，36，81，169，分別是42，62，92，132，而4、6、9、13，後一項跟前項相差「2、3、4」，故下一個應該是跟13相差5，即18和182。偶數項：25，50，100，200是等比數列。

PART

FIVE

Interpretation of Tables and Graphs

Table Chart

Table Chart 1

Expenditures of a Company (in Lakh Rupees) per Annum Over the given Years.

Year	Item of Expenditure				
	Salary	Fuel and Transport	Bonus	Interest on Loans	Taxes
2013	288	98	3.00	23.4	83
2014	342	112	2.52	32.5	108
2015	324	101	3.84	41.6	74
2016	336	133	3.68	36.4	88
2017	420	142	3.96	49.4	98

1. What is the average amount of interest per year which the company had to pay during this period?

A. Rs.32.43 lakhs

B. Rs.33.72 lakhs

C. Rs.34.18 lakhs

D. Rs.36.66 lakhs

Answer: D

Average amount of interest paid by the Company during the given period

$= \text{Rs.}[\frac{(23.4+32.5+41.6+36.4+49.4)}{5}]$ lakhs

$= \text{Rs.}[\frac{183.3}{5}]$ lakhs

= Rs.36.66 lakhs

2. The total amount of bonus paid by the company during the given period is approximately what percent of the total amount of salary paid during this period?

A. 0.1%

B. 0.5%

C. 1%

D. 1.25%

Answer: C

Required percentage $= [\frac{(3.00+2.52+3.84+3.68+3.96)}{(288+342+324+336+420)} \times 100]\%$

$= [\frac{17}{1710} \times 100]\%$

= approximately 1%

3. Total expenditure on all these items in 2013 was approximately what percent of the total expenditure in 2017?

A. 62%

B. 66%

C. 69%

D. 71%

Answer: C

Required percentage = $[\frac{(288+98+3.00+23.4+83)}{(420+142+3.96+49.4+98)} \times 100]\%$

$= [\frac{495.4}{713.36} \times 100]\%$

= approximately 69.45%

4. The total expenditure of the company over these items during the year 2015 is?

A. Rs.544.44 lakhs

B. Rs.501.11 lakhs

C. Rs.446.46 lakhs

D. Rs.478.87 lakhs

Answer: A

Total expenditure of the Company during 2015

= Rs.(324+101+3.84+41.6+74) lakhs

= Rs.544.44 lakhs

5. The ratio between the total expenditure on Taxes for all the years and the total expenditure on Fuel and Transport for all the years respectively is approximately?

A. 4:7

B. 10:13

C. 15:18

D. 5:8

Answer: B

Required ratio

$= [\frac{(83+108+74+88+98)}{(98+112+101+133+142)}]$

$= [\frac{451}{586}]$

= 11.3

= 1013

Table Chart 2

State	Year									
	2013		2014		2015		2016		2017	
	App.	Qual.	App.	Qual.	App.	Qual.	App.	Qual.	App.	Qual.
M	5200	720	8500	980	7400	850	6800	775	9500	1125
N	7500	840	9200	1050	8450	920	9200	980	8800	1020
P	6400	780	8800	1020	7800	890	8750	1010	9750	1250
Q	8100	950	9500	1240	8700	980	9700	1200	8950	995
R	7800	870	7600	940	9800	1350	7600	945	7990	885

1. Total number of candidates qualified from all the states together in 2013 is approximately what percentage of the total number of candidates qualified from all the states together in 2014?

A. 72%

B. 77%

C. 80%

D. 83%

Answer: C

Required percentage

$= [\frac{(720+840+780+950+870)}{(980+1050+1020+1240+940)} \times 100]\%$

$= [\frac{4160}{5230} \times 100]\%$

= 79.54%

= approximately 80%

2. What is the average candidates who appeared from State Q during the given years?

A. 8700

B. 8760

C. 8990

D. 8920

Answer: C

Required average

$= \frac{(8100+9500+8700+9700+8950)}{5}$

$= \frac{44950}{5}$

= 8990

3. **In which of the given years the number of candidates appeared from State P has maximum percentage of qualified candidates?**

A. 2013

B. 2014

C. 2015

D. 2017

Answer: D

The percentages of candidates qualified to candidates appeared from State P during different years are:

For 2013, $(\frac{780}{6400} \times 100)\% = 12.19\%$

For 2014, $(\frac{1020}{8800} \times 100)\% = 11.59\%$

For 2015, $(\frac{890}{7800} \times 100)\% = 11.41\%$

For 2016, $(\frac{1010}{8750} \times 100)\% = 11.54\%$

For 2017, $(\frac{1250}{9750} \times 100)\% = 12.82\%$

Therefore Maximum percentage is for the year 2017.

4. What is the percentage of candidates qualified from State N for all the years together, over the candidates appeared from State N during all the years together?

A. 12.36%

B. 12.16%

C. 11.47%

D. 11.15%

Answer: D

Required percentage = $[\frac{(840+1050+920+980+1020)}{(7500+9200+8450+9200+8800)} \times$ 100]%

= [481043150x 100] %

= 11.15%

5. The percentage of total number of qualified candidates to the total number of appeared candidates among all the five states in 2015 is?

A. 11.49%

B. 11.84%

C. 12.21%

D. 12.57%

Answer: B

Required percentage = $[\frac{(850+920+890+980+1350)}{(7400+8450+7800+8700+9800)} \times 100]\%$

$= [\frac{4990}{42150} \times 100]\%$

$= 11.84\%$

Bar Chart

Bar Chart 1

The bar graph given below shows the sales of books (in thousand number) from six branches of a publishing company during two consecutive years 2016 and 2017.

Sales of Books (in thousand numbers) from Six Branches: B1, B2, B3, B4, B5 and B6 of a publishing Company in 2016 and 2017.

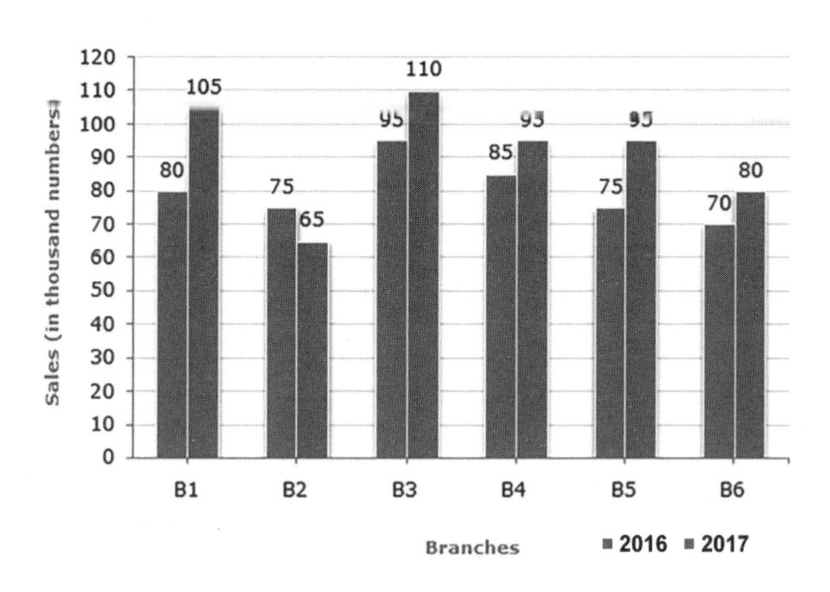

1. What is the ratio of the total sales of branch B2 for both years to the total sales of branch B4 for both years?

A. 2:3

B. 3:5

C. 4:5

D. 7:9

Answer: D

Required ratio = $\frac{(75+65)}{(85+95)} = \frac{140}{180} = 79$

2. Total sales of branch B6 for both the years is what percent of the total sales of branches B3 for both the years?

A. 68.54%

B. 71.11%

C. 73.17%

D. 75.55%

Answer: C

Required percentage = $[\frac{(70+80)}{(95+110)} \times 100]\%$

$= [\frac{150}{205} \times 100]\%$

$= 73.17\%$

3. What percent of the average sales of branches B1, B2 and B3 in 2017 is the average sales of branches B1, B3 and B6 in 2016?

A. 75%

B. 77.5%

C. 82.5%

D. 87.5%

Answer: D

Average sales (in thousand number) of branches B1, B3 and B6 in 2016

$= \frac{1}{3} \times (80+95+70) = (\frac{245}{3})$

Average sales (in thousand number) of branches B1, B2 and B3 in 2001

$= \frac{1}{3} \times (105+65+110) = (\frac{280}{3})$

Therefore, required percentage $= [\frac{(245/3)}{(280/3)} \times 100]\%$

$= (\frac{245}{280} \times 100)\% = 87.5\%$

4. What is the average sales of all the branches (in thousand numbers) for the year 2016?

A. 73

B. 80

C. 83

D. 88

Answer: B

Average sales of all the six branches (in thousand numbers) for the year 2016

$= \frac{1}{6} \times (80+75+95+85+75+70)$

$= 80$

5. Total sales of branches B1, B3 and B5 together for both the years (in thousand numbers) is?

A. 250

B. 310

C. 435

D. 560

Answer: D

Total sales of branches B1, B3 and B5 for both the years (in thousand numbers)

= (80 + 105) + (95 + 110) + (75 + 95)

= 560

Bar Chart 2

The bar graph given below shows the foreign exchange reserves of a country (in million US $) from 2009-2010 to 2017-2017.

Foreign Exchange Reserves Of a Country (in million US $)

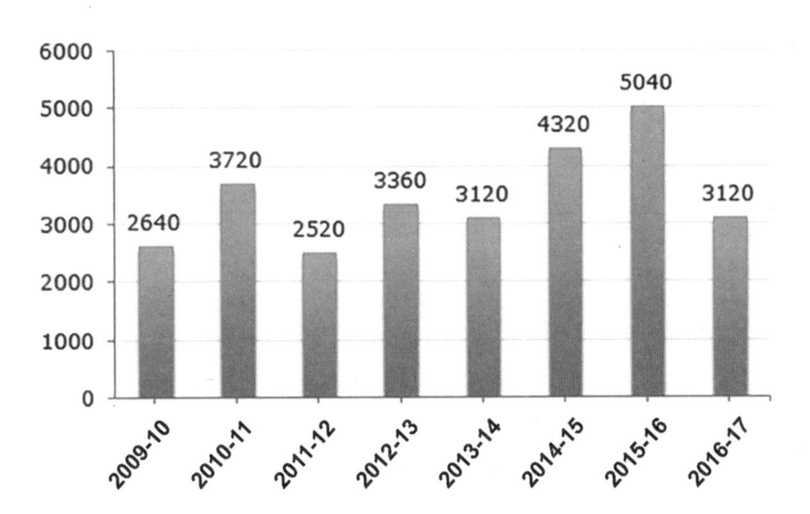

1. The ratio of the number of years, in which the foreign exchange reserves are above the average reserves, to those in which the reserves are below the average reserves is?

A. 2:6

B. 3:4

C. 3:5

D. 4:4

Answer: C

Average foreign exchange reserves over the given period = 3480 million US $.

The country had reserves above 3480 million US $ during the years 2010-11, 2014-15 and 2015-16, i.e., for 3 years and below 3480 million US $ during the years 2009-10, 2011-12, 2012-13, 2013-14 and 2016-17, for 5 yeaRs.Hence, required ratio = 3:5

2. The foreign exchange reserves in 2015-16 was how many times that in 2012-13?

A. 0.7

B. 1.2

C. 1.4

D. 1.5

Answer: D

Required ratio = $\frac{5040}{3360} = 1.5$

3. For which year, the percent increase of foreign exchange reserves over the previous year, is the highest?

A. 2010-11

B. 2011-12

C. 2012-13

D. 2014-15

Answer: A

There is an increase in foreign exchange reserves during the years 2010-11, 2012-13, 2014-15, 2015-16 as compared to previous year (as shown by bar-graph).

The percentage increase in reserves during these years compared to previous year are:

For 2010-11, $[\frac{(3720-2640)}{2640} \times 100]\% = 40.91\%$

For 2012-13, $[\frac{(3360-2520)}{2520} \times 100]\% = 33.33\%$

For 2014-15, $[\frac{(4320-3120)}{3120} \times 100]\% = 38.46\%$

For 2015-16, $[\frac{(5040-4320)}{4320}) \times 100]\% = 16.67\%$

Clearly, the percentage increase over previous year is highest for 2010-11.

4. The foreign exchange reserves in 2014-15 were approximately what percent of the average foreign exchange reserves over the period under review?

A. 95%

B. 110%

C. 115%

D. 125%

Answer: D

Average foreign exchange reserves over the given period

$= [\frac{1}{8} \times (2640+3720+2520+3360+3120+4320+5040+3120)]$ million US $

= 3480 million US $

Foreign exchange reserves in 2014-15 = 4320 million US $.

Therefore, required percentage = ($\frac{4320}{3480}$ x100)% = 124.14%

= approximately 125%

5. What was the percentage increase in the foreign exchange reserves in 2015-16 over 2011-12?

A. 100

B. 150

C. 200

D. 620

Answer: A

Foreign exchange reserves in 2015-16 = 5040 million US $.

Foreign exchange reserves in 2011-12 = 2520 million US $.

Therefore, increase = (5040-2520) = 2520 US $.

Therefore, percentage Increase = ($\frac{2520}{2520}$ x 100)% = 100%

Line Chart

Line Chart 1

Study the following line graph and answer the questions.

Exports from three companies over the years (in Rs.crore)

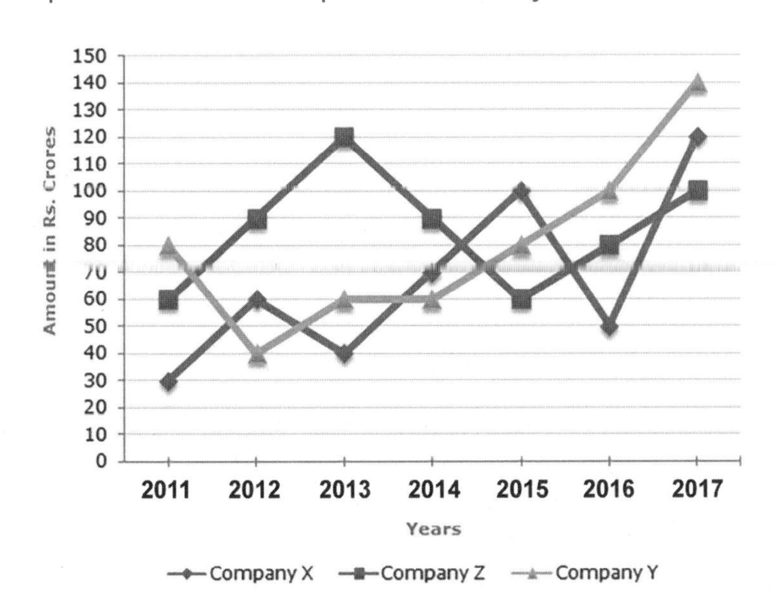

1. For which of the following pairs of years the total exports from the three Companies together are equal?

A. 2013 and 2016

B. 2014 and 2016

C. 2015 and 2016

D. 2013 and 2014

Answer: D

Total exports of the three companies X, Y and Z together, during various years are:

In 2011, Rs.(30+80+60) crores = Rs.170 crores

In 2012, Rs.(60+40+90) crores = Rs.190 crores

In 2013, Rs.(40+60+120) crores = Rs.220 crores

In 2014, Rs.(70+60+90) crores = Rs.220 crores

In 2015, Rs.(100+80+60) crores = Rs.240 crores

In 2016, Rs.(50+100+80) crores = Rs.230 crores

In 2017, Rs.(120+140+100) crores = Rs.360 crores

Clearly, the total exports of the three companies X, Y and Z together are same during the years 2013 and 2014.

2. Average annual exports during the given period for company Y is approximately what percent of the average annual exports for company Z?

A. 87.12%

B. 89.64%

C. 91.21%

D. 93.33%

Answer: D

Analysis of the graph: From the graph it is clear that:

1. The amount of exports of company X (in crore Rs.) in the years 2011-2017 are 30, 60, 40, 70, 100, 50 and 120 respectively.
2. The amount of exports of company Y (in crore Rs.) in the years 2011-2017 are 80, 40, 60, 60, 80, 100 and 140 respectively.
3. The amount of exports of company Z (in crore Rs.) in the years 2011-2017 are 60, 90, 120, 90, 60, 80 and 100 respectively.

Average annual exports (in Rs.crore) of Company Y during the given period

$= \frac{1}{7} \times (80+40+60+60+80+100+140) = \frac{560}{7} = 80$

Average annual exports (in Rs.crore) of Company Z during the given period

$= \frac{1}{7} \times (60+90+120+90+60+80+100) = (\frac{600}{7})$

Therefore, required percentage = $\frac{80}{(600/7)} \times 100]\%$

= approximately 93.33%

3. In which year was the difference between the exports from Companies X and Y the minimum?

A. 2012

B. 2013

C. 2014

D. 2015

Answer: C

The difference between the exports from the Companies X and Y during the various years are:

In 2011, Rs.(80-30) crores = Rs.50 crores.

In 2012, Rs.(60-40) crores = Rs.20 crores.

In 2013, Rs.(60-40) crores = Rs.20 crores.

In 2014, Rs.(70-60) crores = Rs.10 crores.

In 2015, Rs.(100-80) crores = Rs.20 crores.

In 2016, Rs.(100-50) crores = Rs.50 crores.

In 2017, Rs.(140-120) crores = Rs.20 crores.

Clearly, the difference is minimum in the year 2014.

4. What was the difference between the average exports of the three Companies in 2011 and the average exports in 2016?

A. Rs.15.33 crores

B. Rs.18.67 crores

C. Rs.20 crores

D. Rs.22.17 crores

Answer: C

Average exports of the three Companies X, Y and Z in 2011

= Rs.[$\frac{1}{3}$ x (30+80+60)] crores = Rs.($\frac{170}{3}$) crores.

Average exports of the three Companies X, Y and Z in 2016

= Rs.[$\frac{1}{3}$ x (50+100+80)] crores = Rs.($\frac{230}{3}$) crores.

Difference = Rs.[($\frac{230}{3}$) - ($\frac{170}{3}$)] crores

$= Rs.(\frac{60}{3})crores$

= Rs.20 crores

5. In how many of the given years, were the exports from Company Z more than the average annual exports over the given years?

A. 2

B. 3

C. 4

D. 5

Answer: C

Average annual exports of Company Z during the given period

$= (\frac{1}{7}) \times (60+90+120+90+60+80+100)$

$= Rs.(\frac{600}{7})$ crores

= Rs.85.71 crores

From the analysis of graph the exports of Company Z are more than the average annual exports of Company Z (i.e., Rs.85.71 crores) during the years 2012, 2013, 2014 and 2017, i.e., during 4 of the given years.

Line Chart 2

The following line graph gives the ratio of the amounts of imports by a company to the amount of exports from that company over the period from 2011 to 2017.

Ratio of Value of Imports to Exports by a Company Over the Years

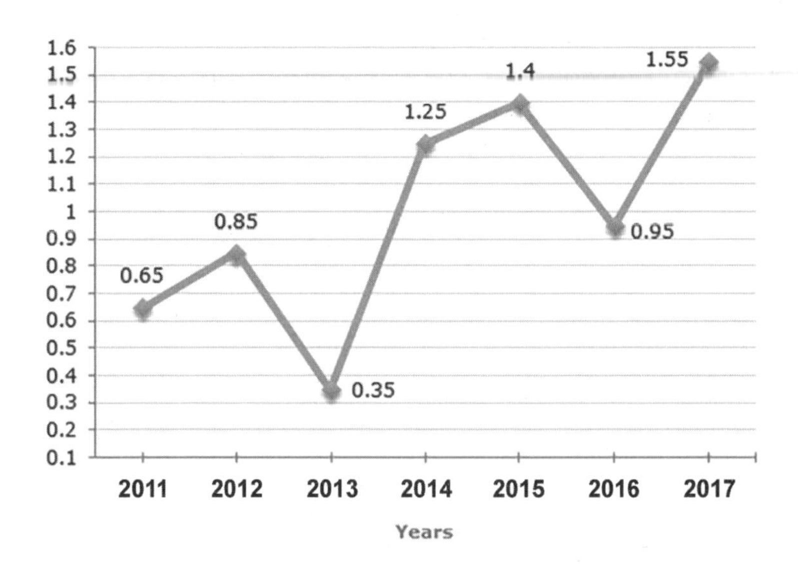

1. If the imports in 2014 was Rs.250 crores and the total exports in the years 2014 and 2015 together was Rs.500 crores, then the imports in 2015 was?

A. Rs.250 crores

B. Rs.300 crores

C. Rs.357 crores

D. Rs.420 crores

Answer: D

The ratio of imports to exports for the years 2014 and 2015 are 1.25 and 1.40 respectively.

Let the exports in the year 2014 = Rs.x crores.

Then, the exports in the year 2015 = Rs. (500-x) crores.

Therefore, $1.25 = (\frac{250}{x})$ $x = (\frac{250}{1.25}) = 200$ [Using ratio for 2014]

Thus, the exports in the year 2015 = Rs.(500-200) crores = Rs.300 crores.

Let the imports in the year 2015 = Rs.y crores.

Then, $1.40 = (\frac{y}{300})$ $y = (300 \times 1.40) = 420$.

Therefore, imports in the year 2015 = Rs.420 crores

2. The imports were minimum proportionate to the exports of the company in the year?

A. 2011

B. 2012

C. 2013

D. 2016

Answer: C

The imports are minimum proportionate to the exports implies that the ratio of the value of imports to exports has the minimum value. Now, this ratio has a minimum value 0.35 in 2013, i.e., the imports are minimum proportionate to the exports in 2013.

3. What was the percentage increase in imports from 2013 to 2014?

A. 72

B. 56

C. 28

D. Data inadequate

Answer: D

The graph gives only the ratio of imports to exports for different yeaRs.To find the percentage increase in imports from 2013 to 2014, we require more details such as the value of imports or exports during these yeaRs.Hence, the data is inadequate to answer this question.

4. If the imports of the company in 2012 was Rs.272 crores, the exports from the company in 2012 was?

A. Rs.370 crores

B. Rs.320 crores

C. Rs.280 crores

D. Rs.275 crores

Answer: B

Ratio of imports to exports in the year 2012 = 0.85

Let the exports in 2012 = Rs.x crores.

Then, $\frac{272}{x} = 0.85 \rightarrow x = \frac{272}{0.85} = 320$

Therefore, exports in 2012 = Rs.320 crores

5. In how many of the given years were the exports more than the imports?

A. 1

B. 2

C. 3

D. 4

Answer: D

The exports are more than the imports imply that the ratio of value of imports to exports is less than 1. Now, this ratio is less than 1 in years 2011, 2012, 2013 and 2016. Thus, there are four such years.

Pie Chart

Pie Chart 1

The following pie-chart shows the percentage distribution of the expenditure incurred in publishing a book. Study the pie-chart and the answer the questions based on it.

Various Expenditures (in percentage) Incurred in Publishing a Book

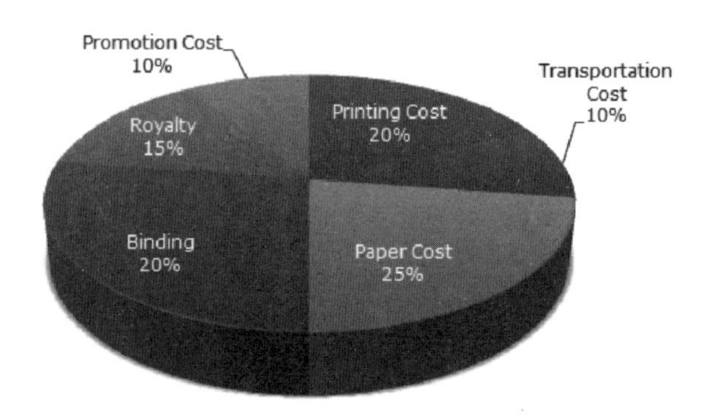

1. If for a certain quantity of books, the publisher has to pay Rs.30,600 as printing cost, then what will be amount of royalty to be paid for these books?

A. Rs.19,450

B. Rs.21,200

C. Rs.22,950

D. Rs.26,150

Answer: C

Let the amount of Royalty to be paid for these books be Rs.r.

Then, 20:15 = 30600:r $\rightarrow$ r = Rs. $\frac{(30600 \times 15)}{20}$ = Rs.22,950

2. What is the central angle of the sector corresponding to the expenditure incurred on Royalty?

A. 15°

B. 24°

C. 54°

D. 48°

Answer: C

Central angle corresponding to Royalty = $(15\% \text{ of } 360)^\circ$

$= (\frac{15}{100} \times 360)^\circ$

$= 54^\circ$

3. The price of the book is marked 20% above the C.P. If the marked price of the book is Rs.180, then what is the cost of the paper used in a single copy of the book?

A. Rs.36

B. Rs.37.50

C. Rs.42

D. Rs.44.25

Answer: B

Clearly, marked price of the book = 120% of C.P.

Also, cost of paper = 25% of C.P

Let the cost of paper for a single book be Rs.n.

Then, 120:25 = 180:n → n = Rs.$\left(\frac{25 \times 180}{120}\right)$ = Rs.37.50

4. If 5500 copies are published and the transportation cost on them amounts to Rs.82500, then what should be the selling price of the book so that the publisher can earn a profit of 25%?

A. Rs.187.50

B. Rs.191.50

C. Rs.175

D. Rs.180

Answer: A

For the publisher to earn a profit of 25%, S.P. = 125% of C.P.

Also Transportation Cost = 10% of C.P.

Let the S.P. of 5500 books be Rs.x.

Then, 10:125 = 82500:x → x = Rs.$\left(\frac{125 \times 82500}{10}\right)$ = Rs.1031250

Therefore, S.P. of one book = Rs.$\left(\frac{1031250}{5500}\right)$ = Rs.187.50

5. Royalty on the book is less than the printing cost by:

A. 5%

B. $33\frac{1}{5}\%$

C. 20%

D. 25%

Answer: D

Printing Cost of book = 20% of C.P.

Royalty on book = 15% of C.P.

Difference = (20% of C.P.) - (15% of C.P) = 5% of C.P.

Therefore, percentage difference

$$= \left(\frac{\text{Difference}}{\text{Printing Cost}} \times 100\right)\%$$

$$= \left(\frac{5\% \text{ of C.P.}}{\text{Printing Cost}} \times 100\right)\%$$

= 25%

Pie Chart 2

The following pie-chart shows the sources of funds to be collected by the National Highways Authority of India (NHAI) for its Phase II projects. Study the pie-chart and answers the question that follow.

Sources of funds to be arranged by NHAI for Phase II projects (in crores Rs.)

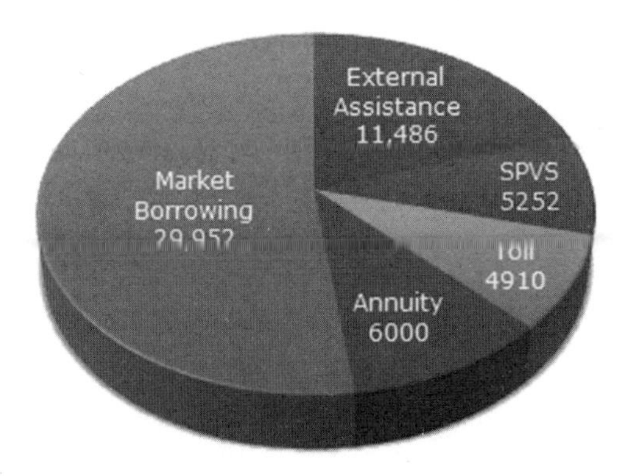

1. Near about 20% of the funds are to be arranged through:

A. SPVS

B. External Assistance

C. Annuity

D. Market Borrowing

Answer: B

20% of the total funds to be arranged = Rs.(20% of 57600) crores

= Rs.11520 crores

= approximately Rs.11486 crores.

Rs.11486 crores is the amount of funds to be arranged through External Assistance.

2. If NHAI could receive a total of Rs.9695 crores as External Assistance, by what percent (approximately) should it increase the Market Borrowing to arrange for the shortage of funds?

A. 4.5%

B. 7.5%

C. 6%

D. 8%

Answer: C

Shortage of funds arranged through External Assistance

= Rs.(11486-9695) crores

= Rs.1791 crores

Therefore, increase required in Market Borrowing = Rs.1791 crores

Percentage increase required = ($\frac{1791}{29952}$ x 100)%

= 5.98%

= approximately 6%

3. If the toll is to be collected through an outsourced agency by allowing a maximum 10% commission, how much amount should be permitted to be collected by the outsourced agency, so that the project is supported with Rs 4910 crores?

A. Rs.6213 crores

B. Rs.5827 crores

C. Rs.5401 crores

D. Rs.5316 crores

Answer: C

Amount permitted

= (Funds required from Toll for projects of Phase II) + (10% of these funds)

= Rs.4910 crores + Rs.(10% of 4910) crores

$= Rs.(4910 + 491)$ crores

$= Rs.5401$ crores.

4. The central angle corresponding to Market Borrowing is

A. 52°

B. 137.8°

C. 187.2°

D. 192.4°

Answer: C

Central angle corresponding to Market Borrowing

$$= \left(\frac{29952}{57600} \times 360^\circ\right)$$

$= 187.2^\circ$

5. The approximate ratio of the funds to be arranged through Toll and that through Market Borrowing is

A. 2:9

B. 1:6

C. 3:11

D. 2:5

Answer: B

Required ratio = $\frac{4910}{29952} = \frac{1}{6.1}$

= approximately $\frac{1}{6}$

看得喜 放不低

創出喜閱新思維

書名　公務員入職　能力傾向測試熱門試題王　第三版
ISBN　978-988-76628-7-7
定價　HK$138
出版日期　2023年3月
作者　Fong Sir
責任編輯　投考公務員系列編輯部
版面設計　Joker
出版　文化會社有限公司
電郵　editor@culturecross.com
網址　www.culturecross.com
發行　聯合新零售（香港）有限公司
地址：香港鰂魚涌英皇道1065號東達中心1304-06室
電話：(852)　2963 5300
傳真：(852)　2565 0919

（作者已盡一切可能以確保所刊載的資料正確及時。資料只供參考用途，讀者也有責任在使用時進一步查證。對於任何資料因錯誤或由此引致的損失，作者和出版社均不會承擔任何責任。）